South Korean Identity and Global Foreign Policy

In the 20th century, South Korea was usually seen as a "shrimp among whales," a minor player with limited agency in regional and global affairs. Korea's risen status as a "middle power" today, however, begs the question about related changes in the South Korean identity or "sense of self" in the world.

In this book, Patrick Flamm presents the first comprehensive and agency-oriented empirical account of South Korean international state identity and Seoul's global foreign policy in the 21st century. Advancing a performative and narrative understanding of identity in International Relations, Flamm uses South Korea's global engagement in peacekeeping and climate diplomacy to offer much-needed insight into the various identity narratives and role conceptions at play. In the case of peacekeeping and climate diplomacy, South Korea's identity as an international actor has been dominated by practices of self-identification that position the country at the brink of advanced countries, aspiring to lead the rest of the world but with the overall objective to maintain national autonomy in a changing regional and global context.

South Korean Identity and Global Foreign Policy is a must-read for scholars of International Relations, Foreign Policy Analysis, and Asian/Korean Studies.

Patrick Flamm is Lecturer in International Relations at the School of History, Philosophy, Political Science and International Relations at Victoria University of Wellington. His research interests are South Korea as a rising power in global affairs, identity and foreign policy, as well as international environmental and Antarctic politics. Patrick studied Political Science and International Relations at Goethe-University Frankfurt, Germany, and Korea University, Republic of Korea, and holds a PhD in Asian Studies from the University of Auckland.

Role Theory and International Relations
Edited by Cameron G. Thies
Arizona State University
Juliet Kaarbo
University of Edinburgh

The *Role Theory and International Relations Series* aspires to attract and publish the latest and best research integrating knowledge in the field of International Relations with role theory. This aspiration cuts across a wide swath of subfields, including foreign policy analysis, peace and security studies, international political economy, diplomatic studies, and international organization. While each of these subfields of study is presently organized as an "island of theory," this series intends to integrate their signature phenomena within a system of knowledge, a "theory complex" or an alliance among different subfields. This series showcases the ability of role theory to generate useful theoretical insights on its own or in combination with existing theories across these traditional subfields. Role theory's conceptual repertoire, plus its ability to span multiple levels of analyses and the major meta-theoretical divides in the discipline position it to be an important integrative force in the study of International Relations.

Domestic Role Contestation, Foreign Policy, and International Relations
Edited by Cristian Cantir and Juliet Kaarbo

Role Theory and Role Conflict in U.S.-Iran Relations
Enemies of Our Own Making
Akan Malici and Stephen G. Walker

South Korean Identity and Global Foreign Policy
Dream of Autonomy
Patrick Flamm

Role Theory in the Middle East and North Africa
Politics, Economics and Identity
Yasemin Akbaba and Özgür Özdamar

For more information about this series, please visit: www.routledge.com/Role-Theory-and-International-Relations/book-series/RTIR

For Katharina und Nikolaus Thun, my dear grandparents

Contents

Figures and Tables

Figures

Tables

Acknowledgments

I would like to thank and acknowledge a few persons and institutions without which this book would not exist: at the University of Auckland, Changzoo Song for his unshakable optimism and trust and Rumi Sakamoto for the enduring encouragement and support. David Hundt at Deakin University and especially Cameron Thies at Arizona State University provided very constructive feedback on my work, which proved invaluable for the revision stage. It was Cameron who suggested I write a book for the *Role Theory and International Relations Series*, of which he is the editor together with Juliet Kaarbo. Thank you all for your immense support and advice. At Routledge, I would like to thank Natalja Mortensen, who was very supportive and always helpful.

During my PhD I was lucky enough to receive a very generous Korean Studies Doctoral Scholarship from the University of Auckland and was also able to spend a few months at the Academy of Korean Studies in Seongnam, South Korea, as a postgraduate fellow. Thank you for this crucial financial support during my studies. Furthermore, I would like to express my gratitude to the Korea Foundation for the one-year postdoctoral fellowship that brought me to Victoria University of Wellington and enabled me to focus on the revision of the book manuscript. At Victoria, I would like to especially thank Xavier Marquez and David Capie, as well as Jonette Crysell, Sam Bigwood and Caroline McDonald, for welcoming me to Wellington and making me a part of the Political Science and International Relations Programme. Thank you for the opportunity to be a part of this fantastic academic and collegial environment.

There are also many friends and fellow travelers on this academic journey whose influence and support I would like to acknowledge here, especially Marcelo Mendes de Souza, Anna Bauer, Simon Opit, Zbigniew Dumieński, Nicholas R. Smith, Rodrigo Ramalho, Alex Payne, and Sidney Smith. This book is dedicated to my dear grandparents, Katharina and Nikolaus Thun. In a way, their journey, which began behind the Iron Curtain in Romania and brought them to Southern Germany, came to an appropriate end with this book written on the other side of the planet in New Zealand. Thank you for everything. I am also very grateful to my parents, Kathy and Peter Flamm. It is only because of their unconditional backing and trust that I was able to go to university in the first place. Finally, I want to thank Pia, who moved to the other end of the world with me, for her unbroken support and understanding.

Note on Transliteration

There are several transliteration systems for Korean, with the so-called McCune–Reischauer and Revised Romanization of Korean schemes being the most widely used. The transliteration of Korean in this study is based on the latter, which was introduced in July 2000 by the South Korean government as the official romanization system for Korean. Transliterations have been done with the help of the Korean Romanization Converter (http://roman.cs.pusan.ac.kr/input_eng.aspx) provided by Busan National University. Furthermore, Korean names are written according to the customary usage in Korean with family names preceding given names. Exemptions are made in cases where a different name order or spelling has already been established in the academic literature, for example, “Syngman Rhee” instead of “Lee Seung-man.”

Abbreviations

AKS	Academy of Korean Studies
BBC	British Broadcasting Corporation
CO_2	Carbon Dioxide
COP	Conference of the Parties
DAC	OECD Developmental Aid Committee
DLP	Democratic Labor Party
DP	Democratic Party
DPRK	Democratic People's Republic of Korea
FPA	Foreign Policy Analysis
G20	Group of Twenty
GCF	Green Climate Fund
GFC	Global Financial Crisis
GGGI	Global Green Growth Institute
GNP	Grand National Party
HLF-4	Fourth High Level Forum on Aid Effectiveness
IMF	International Monetary Fund
INTERFET	International Force for East Timor
IR	International Relations
MIKTA	Mexico, Indonesia, South Korea, Turkey, Australia
MINUSTAH	Mission des Nations Unies pour la stabilisation en Haïti/ United Nations Stabilisation Mission in Haiti
MNF	Multinational Force
NAMAs	Nationally Appropriate Mitigation Actions
NATO	North Atlantic Treaty Organization
NPP	New Progressive Party
NPT	Nuclear Proliferation Treaty
NSS	Nuclear Security Summit
ODA	Official Development Assistance
OECD	Organisation for Economic Co-operation and Development
OPCON	Operational Command
PCGG	Presidential Committee on Green Growth
PKO	Peacekeeping Operation
PRC	People's Republic of China

ROK	Republic of Korea
SINGG	Seoul Initiative Network on Green Growth
SIT	Social Identity Theory
UN	United Nations
UNFCCC	United Nations Framework Convention on Climate Change
UNIFIL	United Nations Interim Force in Lebanon
UNMISS	United Nations Mission in South Sudan
UNOSOM	United Nations Operation in Somalia
UPP	Unified Progressive Party
URL	Uniform Resource Locator
US	United States

1 Introduction

In the 20th century Korea lost its independence, was colonized, then liberated, devastated by war, divided by geopolitics, and ruled by military dictators. Robinson therefore speaks of a "turbulent twentieth century" for Koreans, with a "disproportionately important role in in the last hundred years of world history."[1] The hardships of the Korean 20th century are usually explained with the country's geographical position amid bigger, more powerful neighbors[2] and often described with the metaphor of a small shrimp among big whales, after the Korean proverb "a shrimp's back breaks in a fight among whales."[3] The shrimp metaphor alludes to the country being merely a small nation geographically caught between powerful neighbors, Japan, China, and Russia, and consequently a perennial victim at the hands of greater forces. It is Korea's geopolitical vulnerability that is said to have led to colonialism, war, and national division. As a shrimp, Korea has for a long time been seen as a minor player with limited agency that was determined by an unfavorable regional power structure and thus in need of protecting allies, whether these were the Chinese dynastic emperors in a Sino-centric world of tributary relations, Imperial Japan amid Western encroachment on Asia, or the US during times of national division and the Cold War.[4]

But Koreans in the Southern Republic of Korea (ROK or South Korea hereafter) have also experienced unprecedented economic development, built a stable democracy, and initiated (a once-promising) rapprochement and engagement with their estranged Northern cousins through what became known as the "sunshine policy." With regard to the achievements of South Korea, in contrast, Cumings speaks metaphorically of Korea entering the world stage and finding its "place in the sun"[5] as one of only a handful in the "solar system of advanced industrial states."[6] This "rising Korea"[7] is generally referred to as a middle power that has been repeatedly "punching above its weight"[8] or is described metaphorically as a "dolphin"[9] that is outsmarting[10] other sea creatures. These images refer to the facts that Seoul now is part of the G20, the Organisation for Economic Co-operation and Development (OECD), the "club of rich countries," and the OECD Developmental Aid Committee (DAC), as the first former aid recipient country ever to do so. Just a few decades after the Korean War, in which soldiers from over 20 United Nations (UN) member states fought alongside South Koreans, today ROK troops are serving as UN peacekeepers in South Sudan and in the coastal waters of Lebanon. The country also initiated its own space program[11] and is an emerging power in polar politics.[12] Therefore, at the dawn of the 21st century, South Korea started to be seen as "no longer a pawn but a pivotal player in Northeast Asian affairs"[13] and in international politics.

Today, international scholars and pundits project certain hopes and expectations on the country for the present and the future. For Ikenberry and Mo, the middle-power ROK is already at the center of today's global order and will be one of the key states for the future of liberal internationalism,[14] quite in contrast to other rising powers, such as China or Russia. Snyder, however, sees Seoul more through the prism of its long-standing alliance with the US and thus as a supporter of US hegemony in East Asia, as well as a key pillar of Washington's pivot to Asia.[15] The story told is that of Korea's journey from an "insular country to an outward-looking nation,"[16] that engages with the rest of the world to an unprecedented extent.[17] Typical for these perspectives is the realist focus on South Korea's position or standing vis-à-vis other states, which is something that can be assessed from the outside without necessarily consulting Korean domestic perspectives. To these observers, rising South Korea looks like a middle power, walks like a middle power, and thus is a middle power.[18] In their prognoses, South Korea is likely to and is also well advised to proceed on a middle-power trajectory toward a stronger US alliance, more business with China, further contributions to the common good, eventual unification with the North, and maybe even a "normal" relationship with the former colonial master, Japan.[19] What is missing from this picture, however, is how the ROK and South Koreans themselves are seeing their role and place in the world. What is missing is how South Korea as a country relates itself to a globalized world. This crucial question is more about self-identity and agency than structural position and behavior.

To be clear, identity is a broad, multilayered, and multifaceted concept and can be explored among individuals, social groups, or polities, and South Korea's international rise and encounter with a globalized world have been investigated on different levels of analysis and from different angles. In the cultural and everyday social spheres, for example, Lee explores how national and global identities were negotiated on Korea's theater stages,[20] while Chun and Han researched the global ambitions of language-traveling Korean youth and how these students saw their place in the international order compared to Canadians and Filipinos.[21] In the religious and political spheres, Han documents how Korean evangelical missionaries "constructed and presented Korean history as a teleology of modernization, industrialization, and Christianization"[22] in Africa, and Robertson studied the diplomatic style of Seoul's emissaries on the world stage, finding, for example, "a preoccupation with status, widening generational change, [and] high levels of cosmopolitanism"[23] among South Korean diplomats.

There is no comprehensive study as yet looking at the formation and changes of state identity from the perspective of South Korean global foreign policy, neither in the International Relations nor Korean Studies literature, which is arguably still one of the main avenues for the construction and formation of South Korean international identity.[24] A telling example of this gap is the edited volume *Korea Confronts Globalization*,[25] which features chapters on globalization and women's identities, regionalism, labor relations, democratization, party politics, and religions, but none on Korea's relations to other countries or global governance. What is most significant, however, is that a sound understanding of a state's identity is, first, instrumental in understanding the stability and trajectory of its foreign policy behavior. Identity matters, as constructivism teaches us, because "without interests identities have no motivational forces, without identities interests have no direction."[26] Second, taking the self-identity of a state actor seriously is a way of taking the state actor's own agency seriously, especially in a postcolonial context like this. Simply inferring the identity of a polity from its behavior, or its role performance, toward other political communities ends up as a convenient ignorance that obscures a better understanding of a polity and its constitutive inner conflicts and deviant voices.

The Research Question of the Book

Against this background, a number of questions arise: How did the proverbial shrimp grow into a dolphin? What happened to the old shrimp self-understandings of postcolonialism, anticommunism, and anti-imperialism? Is the wishful optimism about Korea's future global foreign policy behavior by some observers justified, and is the rising Korea really going to be a cornerstone of liberal internationalism? Is the global outreach of the early 21st-century South Korea commonplace among foreign policy makers, or is it politically contested within the country? Was it maybe just a phase with a "globalist school"[27] at the helm of Seoul's foreign policy agenda, or has there been a substantial and lasting change in state identity supported by the foreign policy community and the Korean society at large? And what is the self-understanding put forward by this globalist school? Is it about transcending national boundaries, as Robertson's finding of "high levels of cosmopolitanism"[28] among South Korean diplomats might indicate, or a mere extension of the "national interest" into the global arena? All these questions, however, ultimately come down to the main question of how does rising South Korea's global foreign policy and Seoul's unprecedented international leadership ambitions relate to a Korean collective sense of self or the country's international identity?

Generally, the narrative from "insular country to an outward-looking nation"[29] that focuses on the hermit kingdom turned global model student should be treated with caution. There are three main reasons for this. First, when South Korea opened its economy for global markets in the 1990s, it was a "state-enhancing, top-down strategic plan"[30] termed *segyehwa*, literally meaning globalization, initiated by President Kim Young-sam (1993–1998) and aimed at enhancing the nation's international competitiveness. The plan was to restructure Korea not just economically but also politically, socially, and culturally in an attempt to transform Korea into an "advanced nation." For several critics of the ultimately unsuccessful program, *segyehwa* remained "subservient to nationalist goals,"[31] with "no fundamental learning—no paradigm shift [. . . and] only situation-specific tactical adaptation."[32] Others suggested that globalization, by being "appropriated by the Korean nation,"[33] has actually been reinforcing nationalism, leading to a "paradox"[34] of Korean globalization. To illustrate this, a recent media analysis of the South Korean global citizenship discourse of the last two decades also supports a skeptical observer in finding that "global citizenship is primarily about national advancement rather than any sort of transcendence of national identity into cosmopolitan ideals."[35]

Second, the entrenched "generally negative, self-doubting narrative"[36] that Korea is a small country and a perennial victim of larger forces surely must remain influential, but how so? The experiences of Western encroachment, colonization, and national division have fostered a deep sense of the inherent value of national sovereignty in both South and North Korea. What happened to those traditional identity understandings that have been politically relevant for decades? True, one segment of Korean society, the younger generation, is beginning to see things differently. According to the findings of a long-term study, "pride in ROK's achievement and its growing international importance is a key part of [younger Koreans'] national identity,"[37] but how this relates to changing concrete foreign policy actions remains unclear, especially because Koreans in their 20s have been found to be "security conservative," often in accordance with the generations older than 60 years.[38]

Third, one may want to challenge the purported novelty of the narrative about Korea's rise. Even Cumings, in his seminal monography *Korea's Place in the Sun*, not only referred the metaphor to Korea's 20th-century development but had already titled one of the book's subchapters on the scientific, technological, and cultural advances of Goryeo (918–1392) and medieval Joseon Korea "Korea in the Sun."[39] For example, he refers to Koreans inventing movable metal type printing in 1234, long before Johannes Gutenberg in Europe. The capital of Goryeo, Kaesong, back then was an important trade hub at the end of the Silk Road, with traders from as far as Arab lands, and was allegedly the most populous city in the whole region.[40] Other culturally important products of the era include the refined jade-green Celadon porcelain as well as the World Heritage Tripitaka Koreana, a comprehensive Buddhist canon on woodblocks. Furthermore, there are contributions to astronomy, agronomy, and, most important, the invention of the scientific written Korean alphabet, *Hangeul*, introduced by King Sejong in 1446.[41] Back then, Japan was considered a mere recipient of culture and an inferior, peripheral country, far from the center and thus the dignifying influence of the Chinese emperor, the Son of Heaven.

This worldview is also mirrored in the *Gangnido* world map of 1402, one of the oldest surviving Northeast Asian world maps, where an enlarged Korea sits just next to the central China, connected to the Middle East, and a tiny continental appendix in the West, Europe. This is why Cumings states that Koreans have always seen their country as "important, advanced, significant,"[42] and as a nation that is now, after the humiliation of the imperial era, "reoccupying its appropriate position in the world."[43] This perspective strikingly mirrors those of modern Chinese and Indian status aspirations in today's international system and begs the question about the country's own understanding of its appropriate position in the world. A look at how Koreans themselves calibrate and reflect on their role in the world seems more than warranted, especially so as the South Korean debate is still ongoing:

> South Korea, which for many years lacked a forward-looking strategy, is now in a position to opt for something longer term and ambitious. These days, there are growing debates and burgeoning discourses about the desirable future trajectory of South Korean foreign policy.[44]

South Korean identity is certainly "transforming"[45] or "at crossroads,"[46] but how so and in which direction? And most important, what does it mean for South Korean international state identity and its future trajectory as an international actor?

Accordingly, this study investigates the relationship between the global foreign policy and international state identity of 21st-century South Korea as its main question. There are a number of subquestions, however: first, it has to be asked whether a "globalist" self-understanding that aims "to rethink the whole paradigm of calculating national interests and to broaden the range of South Korean's influence"[47] in a "post-Westphalian global order"[48] is firmly established in the domestic foreign policy discourse. Furthermore, it has to be seen whether a globalist outlook actually translates into a normative post-Westphalian "globalist" identity that embraces international law and limitations of national sovereignty and adheres to international responsibilities as a "global citizen." Only then the hopes that Korean middle powermanship lead to a brighter future of liberal internationalism would seem realistic and justified. The second subquestion asks about the alleged established postcolonial norms that prioritize national sovereignty and anti-imperialist justice. How do these identity narratives relate to South Korean global foreign policy in the 21st century? The third subquestion aims at the potential influence of pan-Asian narratives that have been popular in the late 19th and 20th centuries, when the perceived threat of Western internationalization was confronted by pan-Asian cooperation or, for example, with an Asian values debate.

The Argument of the Book

Before one can attempt to answer these questions, given the aforementioned "definitional anarchy,"[49] identity as an analytical concept has to be briefly defined. Identity is one of the main concepts in International Relations (IR) and its prominence has much to do with the emergence of the interpretative and postrationalist approach of constructivism since the 1990s. However, due to its inherent complexity, there have been a great diversity of different ontological and epistemological assumptions employed in the study of identity, with recent attempts to coordinate and integrate different identity conceptualizations in order to realize the potential of identity-based research while recognizing the plurality of identity perspectives.

Acknowledging the complexity of identity, which encompasses the individual and the collective, state and nation, and historical, cultural, and material aspects, in effect, forces the rebuttal of simple or parsimonious conceptualizations of identity. There is a commonsense agreement, however, that having an identity means having a sense of self, without which the actor would be unable to perceive and enact any meaningful agency, a precarious state that has been termed "ontological insecurity." The challenge for IR constructivism, in particular, is how to understand the formation of this inherently dynamic and situational idea about oneself for organized collective entities. Any investigation into a state's identity has to make clear the specific premises and parameters applied: What is the specific operationalization, the level of analysis, whose identity is investigated, where actorhood is seen as located; who is acting on behalf of the identity bearer; and how is identity formed and changed?

If one is particularly interested in the identity representations underpinning foreign policy practice of a given state as in the case of this book, it seems important to only claim inquiry into the "international identity" of that state, and not the "national identity" or just "identity" as the broader culture of that community as a whole, even though these are certainly connected. This is why the "nation" is not a fitting referent object for identity in this case, especially as there are two Korean states as different manifestations of Korean "nation-stateness." Furthermore, Korea's "international identity" is not solely determined by Korean states but, for example, also by Koreans abroad, whether they are diplomats, diasporic communities, travelers, or sports teams competing in international tournaments. Last, in other contexts, "state identity" may mean the relationship between the state (the government) and societal/substate actors, such as trade unions or separatist movements. In consequence, what is in the focus of the book here is the international identity of the ROK state in relation to Seoul's foreign policy.

Furthermore, this book argues that one solution to the identity problematique lies in a multidimensional understanding of identity as a narratively and performatively (as well as emotionally) formed phenomenon. Here, the rich conceptual language and multilevel perspective of the role-theoretical approaches can have an integrative function regarding the clarification of terminology and the opening of an empirically promising way of grasping "international state identities" beyond their manifestations in social identities. This seems significant given that an actor's identity is more than just its social manifestation in role identities and thus more than "something negotiable between states,"[50] as Zehfuss puts it. Very simply, roles are the results of the encounter of an actor's identity with the expectations of a given social environment. Any role has to be based on a corresponding identity element, and any identity has to be reaffirmed by respective role-playing. Hence, identity affirmation is only possible due to role enactment, as only social roles give identity meaning.[51]

The symbolic interactionist variant of role theory, especially, which is based on the Meadian "I and Me" and focuses on the interpretative process of self-identification, lends itself to the exploration of agency-oriented identity issues related to foreign policy. A sense of self or a self-identity is here acquired through the reflection of and interaction with significant and generalized Others. Theoretically, this sense of self is further expressed and habitualized in a narrative way, and with the focus on role performance, role theory is very much in line with research approaches focusing on the performative and situational aspects of identity as well. Furthermore, unpacking the identity formation within the "ego" opens the future potential to also trace the emotional dimension of identification politics along the horizontal and vertical contestation of the formation of role conceptions, even though this is not part of this study.

Based on these symbolic interactionist role-theoretical foundations, this study presents a theoretically rigorous and empirically rich approach for the inquiry into state identity. Here, the nation-state is not treated as a unitary actor but as a social structure itself within which different domestic political actors compete for dominance. In the conceptualization of state identity developed here, there are two observable layers of the international self-identity of state actors. These are the layer of concretely manifested role identities in a given social context and the more basic layer of self-images that have been employed for the legitimation of specific role conception and related actions. In order to understand and not just describe a country as an international actor, it is necessary to focus on how these role identities have been legitimated with reference to self-images within the domestic decision-making process.

An inductive, textual analysis of domestic role contestation in these respective foreign policy discourses can account for the different self-identifications at play in different foreign policy decisions, while the focus lies on the contestation and stability of these self-understandings over time. The comparison of these retrieved self-images in these self-identifications from different policy fields and over time allows inference of larger patterns of international self-identity narratives. As snapshots of an international, collective sense of self, these self-images enable insights into the larger international identity dynamics behind Korea's global foreign policy. This book applies this approach to two different foreign policy fields in two separate case studies on South Korea's international peacekeeping and climate diplomacy, as these foreign policy fields have been identified as some of the most salient for South Korea's emergence as a global power in the respective literature. In these cases, South Korea is not only expressing global leadership claims but is also, at the same time, very much exposed to international expectations about their foreign policy behavior.

Thus, in the approach presented here, role identities really are only the outer layer of South Korean international state identity, with the inner layer consisting of self-images on which different role conceptions and eventually established role identities are built on. These self-images that are utilized in the legitimation of specific foreign policy actions are treated as snapshots of the elusive broader international identity of South Korea. Accordingly, the argument of this book here is that only the comprehensive mapping of these different self-identifications—on both layers of role identities/conceptions and self-images—deserves to be called an acceptable assessment of a state's international identity, because such a description is reflecting their contestation and stability over different foreign policy fields and over time.

Given the application of this developed approach to the investigation into South Korea's international state identity then, what are the empirical findings of this study? Among the retrieved role conceptions from the two case studies on South Korean international peacekeeping and climate diplomacy, one notices the presence of seemingly normatively globalist, post-Westphalian role conceptions, such as "global model nation" and "responsible international member," as well as rhetoric references to global morality or international interdependence. A closer look at the discursive linkages to the more basic level of self-images, however, leads to more traditional narratives of legitimation that prioritize sovereign autonomy and national advancement as key objectives of South Korea's global foreign policy. The main self-images are that of a "rising Korea," focusing on Korea's elevated status as a result of economic development and democratization, on which the understanding of a "leading Korea" builds with a pronounced and confident leadership claim. Apart from these major self-images, there are slightly deviating and more marginal self-understandings, for example, where Korea's development is not yet successfully finished and must still be safeguarded or the image of Korea as an inherently peaceful nation that does not participate in modern-day resource "imperialism" practiced by the US and China, as well as Japan.

It is because of the absence of a pattern of self-identification that is transcending national boundaries with global commons oriented or post-Westphalian globalist inclination, and because of a strong focus on international standing and autonomy in the self-identification practices in the observed cases, that this book concludes that South Korean international state identity is dominated by the objective to maintain national autonomy above all in a changing regional and global context: global foreign policy has been a means of validating the risen international status of the country and a way of trying to establish South Korea as a rule maker rather than a rule follower. Its legitimation was dominated by narratives about fostering a premium nation brand and extensive ties around the world, against a referenced historiography of vulnerability due to the lack of international connections and knowledge. Global foreign policy here is quite explicitly seen as a realization of a dream of greater autonomy in a challenging regional security environment.

The Plan of the Book

The rest of this book is divided into three parts. The first part sets the stage for the inquiry into South Korean state identity and foreign policy. First, the literature on Korea's relations with the outside world, its foreign policy, and its self-understanding are reviewed in Chapter 2. As a consequence, we will arrive at an empirical and conceptual double gap: empirically, we are still missing a comprehensive assessment of South Korean state identity as a rising global power while conceptually there is a lack of agency-oriented approaches to studying the state identity of South Korea as an international actor. Second, the IR literature on identity as an analytical concept is reviewed in Chapter 3. Because of the elusiveness of the concept and the "definitional anarchy"[52] still prevalent in IR identity studies, the integrative potential of symbolic interactionist role theory for the inquiry into international state identities is introduced.

In the second part, the investigation into Korea's role-playing on the global stage takes place. First, the methodological approach inspired by symbolic interactionist role theory for the investigation of international state identity is presented in Chapter 4. The approach is informed by theoretical insights from IR constructivism, International Political Sociology, and role theory as developed in Foreign Policy Analysis (FPA), for which "[s]tates [. . .] hold a host of national and international roles that constitute their identity, regulate their behavior and shape the international order."[53] By analyzing the foreign policy discourse of domestic political legitimation for foreign policy actions in the search for self-identification practices, which are understood as linkages between role conceptions and self-images, we will arrive at a comprehensive account of South Korean state identity in relation to foreign policy, with the self-images even allowing insights into larger South Korean international identity. An understanding of these two layers of international state identity, the easily observable manifestation of dominant role conceptions as role identities and the directly related but more elusive self-images, provides a better and agency-oriented insight into the constitution of South Korea as an international actor than do conventional studies. This approach will be applied in two case studies on South Korean international peacekeeping as well as international climate diplomacy in Chapters 5 and 6.

In the third and last part of the book, a comprehensive account of the self-identification practices retrieved from the two case studies will help to answer the study's main research question about the relationship between South Korea's international state identity and global foreign policy in the early 21st century. Furthermore, the key argument of this book, that South Korean international state identity is mostly dominated by the objective to raise the country's international standing in order to foster national autonomy in a changing regional and global context, is presented in Chapter 7. Last, there are concluding remarks about the limitations of this work and further potential research questions in Chapter 8. This study aims to be a theoretically sound and empirically rich agency-oriented inquiry into the international state identity of Korea as an emerging global power. The ultimate goal of this book is to provide a better and more comprehensive understanding of the international actor South Korea while combining an IR Theory framework with Korean Studies expertise in language, history, and culture.

Notes

1. Robinson, Michael E. 2007: *Korea's Twentieth-Century Odyssey: A Short History*, Honolulu: University of Hawaii Press, p. 1.
2. Oberdorfer, Don 1997: *The Two Koreas: A Contemporary History*, Reading, MA: Addison-Wesley, p. 9; Kim, Samuel S. 2006: *The Two Koreas and the Great Powers*, Cambridge: Cambridge University Press, p. 1; Chun, Chaesung 2013: "South Korea's Foreign Policy and East Asia", in: Rüdiger Frank and John Swenson-Wright (eds.), *Korea and East Asia: The Stony Road to Collective Security*, Leiden: Brill, p. 162; De Ceuster, Koen 2005: "Pride and Prejudice in South Korea's Foreign Policy", *Copenhagen Journal of Asian Studies*, Vol. 21, pp. 64–90.
3. In Korean, "*Goraessaume saeu deung teojinda*."
4. Compare Bae, Kichan 2007: *Korea at the Crossroads: The History and Future of East Asia*, Seoul: Happy Readings Books.
5. Cumings, Bruce 2005: *Korea's Place in the Sun: A Modern History*, New York: W. W. Norton Company.
6. Ibid.: p. 9.
7. Hwang, Balbina Y. 2004: "Introduction: The Rise of Korea", *East Asia*, Vol. 21, No. 2, pp. 39–41.
8. Bradford, Colin I. 2015: "South Korea as a Middle Power in Global Governance: 'Punching Above Its Weight' Based on National Assets and Dynamic Trajectory", in: Scott A. Snyder (ed.), *Middle-Power Korea: Contributions to the Global Agenda*, New York: Council on Foreign Relations Press, pp. 8–20.
9. Oh, Kongdan 1999: "South Korea's Foreign Policy: A Dolphin among Whales?", *Paper presented to the 14th Annual Conference of the Council on US-Korean Security Studies*, 27–30 October, Arlington, VA; Brazinsky, Gregg 2007: *Nation Building in South Korea: Koreans, Americans, and the Making of a Democracy*, Chapel Hill: University of North Carolina Press; Tudor, Daniel 2014: "Korea Is a Dolphin, Not a Shrimp", *Korea Joongang Daily*, 3 February. Available at: http://mengnews.joins.com/view.aspx?aId=2984326 [Accessed 5 November 2018].
10. Brazinsky, G. 2007: p. 7.
11. Kang, Tae-jun 2015: "South Korea's Quest to Be a Major Space Power", *The Diplomat*, 27 March. Available at: http://thediplomat.com/2015/03/south-koreas-quest-to-be-a-major-space-power/ [Accessed 5 November 2018].
12. Brady, Anne-Marie and Kim, Seungryeol 2013: "Cool Korea: Korea's Growing Antarctic Interests", in: Anne-Marie Brady (ed.), *The Emerging Politics of Antarctica*, London/New York: Routledge, pp. 75–95; Watson, Iain 2016: "From Middle Power to Pivot Power: Korea as an Arctic Observer in the Age of Eurasia", *Pacific Focus*, Vol. 31, No. 3, pp. 333–356.
13. Kim, S.S. 2006: p. xiv.
14. Ikenberry, G. John and Mo, Jongryn 2013: *The Rise of Korean Leadership, Emerging Powers and International Order*, New York: Palgrave Macmillan, pp. 164–165.
15. Snyder, Scott A. 2018: *South Korea at the Crossroads: Autonomy and Alliance in an Era of Rival Powers*, New York: Columbia University Press.
16. Bradford, C.I. 2015: p. 11.
17. Armstrong, Charles K. 2014: *The Koreas*, London/New York: Routledge, p. 34.
18. See also O'Neill, Andrew 2015: "South Korea as a Middle Power: Global Ambitions and Looming Challenges", in: Scott A. Snyder (ed.), *Middle-Power Korea: Contributions to the Global Agenda*, New York: Council on Foreign Relations Press, p. 78ff.
19. Compare Kim, Hyun-Wook 2011: "Obama's East Asia Policy and US-South Korea Relations", in: Melissa H. Conley Tyler and Wilhelm Hofmeister (eds.), *Going Global: Australia, Brazil, Indonesia, South Korea and South Africa in International Affairs*, Singapore: Konrad Adenauer Stiftung/Deakin, ACT: Australian Institute of International Affairs, p. 73; Chun, C.S. 2013: pp. 167–168.

20. Lee, Hyunjung 2015: *Performing the Nation in Global Korea: Transnational Theatre*, London: Palgrave Macmillan.
21. Chun, Jennifer Jihye and Han, Ju Hui Judy 2015: "Language Travels and Global Aspirations of Korean Youth", *Positions: East Asia Cultures Critique*, Vol. 23, No. 3, pp. 565–593.
22. Han, Ju Hui Judy 2011: "'If You Don't Work, You Don't Eat': Evangelizing Development in Africa", in: Jesook Song (ed.), *New Millennium South Korea: Neoliberal Capital and Transnational Movements*, London/New York: Routledge, p. 154.
23. Robertson, Jeffrey 2016: *Diplomatic Style and Foreign Policy: A Case Study of South Korea*, London/New York: Routledge, p. 180.
24. See also Berenskoetter, Felix 2014: "Parameters of a 'National Biography'", *European Journal of International Relations*, Vol. 20, No. 1, p. 283; compare Campbell, David 1992: *Writing Security: United States Foreign Policy and the Politics of Identity*, Manchester: Manchester University Press.
25. Chang, Yun-Shik, Seok, Hyun-Ho and Baker, Donald L. (eds.) 2009: *Korea Confronts Globalization*, London/New York: Routledge.
26. Wendt, Alexander 1999: *Social Theory of International Politics*, Cambridge: Cambridge University Press, p. 231.
27. Chun, C.S. 2013: p. 164.
28. Robertson, J. 2016: p. 180.
29. Bradford, C.I. 2015: p. 11.
30. Kim, Samuel S. 2000a: "Korea and Globalization (Segyehwa): A Framework for Analysis", in: Samuel S. Kim (ed.), *Korea's Globalization*, Cambridge: Cambridge University Press, p. 3.
31. Shin, Gi-Wook and Choi, Joon Nak 2009: "Paradox or Paradigm? Making Sense of Korean Globalization", in: Chang Yun-shik, Hyun-Ho Seok and Donald L. Baker (eds.), *Korea Confronts Globalization*, London/New York: Routledge, p. 252.
32. Kim, Samuel S. 2000b: "Korea's Segyehwa Drive: Promise versus Performance", in: Samuel S. Kim (ed.), *Korea's Globalization*, Cambridge: Cambridge University Press, p. 275.
33. Shin, G.W. and Choi, J.N. 2009: p. 265.
34. Ibid.
35. Schattle, Hans 2014: "Global Citizenship as a National Project: The Evolution of Segye Shimin in South Korean Public Discourse", *Citizenship Studies*, Vol. 19, No. 1, p. 54.
36. Tudor, D. 2014.
37. Campbell, Emma 2011: "Changing South Korea: Issues of Identity and Reunification in Formulating the Australia-Korea Security Policy, Foreign Policy, and Wider Relationship", *Korea Observer*, Vol. 42, No. 1, p. 128.
38. Denney, S. and Friedhoff, K. 2013.
39. See Cumings, B. 2005: p. 67.
40. Jun, Seong Ho 2014: "Monetary Authority Independence and Stability in Medieval Korea: The Koryŏ Monetary System through Four Centuries of East Asian Transformation, 918–1392", *Financial History Review*, Vol. 21, No. 3, p. 264.
41. Cumings, B. 2005: p. 67.
42. Ibid.: pp. 9–10.
43. Ibid.: p. 10.
44. Chun, C.S. 2013: p. 155.
45. Ikenberry, G.J. and Mo, JR. 2013: p. 164; Denney, S. and Friedhoff, K. 2013.
46. Rozman, Gilbert 2012: "South Korean National Identity: A Six-Dimensional Analysis", in: Gilbert Rozman (ed.), *East Asian National Identities: Common Roots and Chinese Exceptionalism*, Stanford: Stanford University Press, p. 67; Compare Snyder, S. 2018.
47. Chun, C.S. 2013: p. 164.
48. Ibid.

49. Abdelal, Rawi, Herrera, Yoshiko M., Johnston, Alastair Iain and McDermott, Rose 2006: "Identity as a Variable", *American Review of Political Science*, Vol. 4, No. 4, p. 695.
50. Zehfuss, Maja 2001: "Constructivism and Identity: A Dangerous Liaison", *European Journal of International Relations*, Vol. 7, No. 3, p. 335.
51. McCourt, David M. 2011: "Role-Playing and Identity Affirmation in International Politics: Britain's Reinvasion of the Falklands, 1982", *Review of International Studies*, Vol. 37, No. 4, p. 1619.
52. Abdelal, R. et al. 2006: p. 695.
53. Harnisch, Sebastian, Frank, Cornelia and Maull, Hans W. 2011: "Conclusion: Role Theory, Role Change, and the International Social Order", in: Sebastian Harnisch, Cornelia Frank and Hans W. Maull (eds.), *Role Theory in International Relations: Approaches and Analyses*, London/New York: Routledge, p. 252.

References

Abdelal, Rawi, Herrera, Yoshiko M., Johnston, Alastair Iain and McDermott, Rose 2006: "Identity as a Variable", *American Review of Political Science*, Vol. 4, No. 4, pp. 695–711.

Armstrong, Charles K. 2014: *The Koreas*, London/New York: Routledge.

Bae, Kichan 2007: *Korea at the Crossroads: The History and Future of East Asia*, Seoul: Happy Readings Books.

Berenskoetter, Felix 2014: "Parameters of a 'National Biography'", *European Journal of International Relations*, Vol. 20, No. 1, pp. 262–288.

Bradford, Colin I. 2015: "South Korea as a Middle Power in Global Governance: 'Punching Above Its Weight' Based on National Assets and Dynamic Trajectory", in: Scott A. Snyder (ed.), *Middle-Power Korea: Contributions to the Global Agenda*, New York: Council on Foreign Relations Press, pp. 8–20.

Brady, Anne-Marie and Kim, Seungryeol 2013: "Cool Korea: Korea's Growing Antarctic Interests", in: Anne-Marie Brady (ed.), *The Emerging Politics of Antarctica*, London/New York: Routledge, pp. 75–95.

Brazinsky, Gregg 2007: *Nation Building in South Korea: Koreans, Americans, and the Making of a Democracy*, Chapel Hill: University of North Carolina Press.

Campbell, David 1992: *Writing Security: United States Foreign Policy and the Politics of Identity*, Manchester: Manchester University Press.

Campbell, Emma 2011: "Changing South Korea: Issues of Identity and Reunification in Formulating the Australia-Korea Security Policy, Foreign Policy, and Wider Relationship", *Korea Observer*, Vol. 42, No. 1, pp. 117–144.

Chang, Yun-Shik, Seok, Hyun-Ho and Baker, Donald L. (eds.) 2009: *Korea Confronts Globalization*, London/New York: Routledge.

Chun, Chaesung 2013: "South Korea's Foreign Policy and East Asia", in: Rüdiger Frank and John Swenson-Wright (eds.), *Korea and East Asia: The Stony Road to Collective Security*, Leiden: Brill, pp. 155–180.

Chun, Jennifer Jihye and Han, Ju Hui Judy 2015: "Language Travels and Global Aspirations of Korean Youth", *Positions: East Asia Cultures Critique*, Vol. 23, No. 3, pp. 565–593.

Cumings, Bruce 2005: *Korea's Place in the Sun: A Modern History*, New York: W. W. Norton Company.

De Ceuster, Koen 2005: "Pride and Prejudice in South Korea's Foreign Policy", *Copenhagen Journal of Asian Studies*, Vol. 21, pp. 64–90.

Han, Ju Hui Judy 2011: "'If You Don't Work, You Don't Eat': Evangelizing Development in Africa", in: Jesook Song (ed.), *New Millennium South Korea: Neoliberal Capital and Transnational Movements*, London/New York: Routledge, pp. 142–158.

Harnisch, Sebastian, Frank, Cornelia and Maull, Hans W. 2011: "Conclusion: Role Theory, Role Change, and the International Social Order", in: Sebastian Harnisch, Cornelia Frank and Hans W. Maull (eds.), *Role Theory in International Relations: Approaches and Analyses*, London/New York: Routledge, pp. 252–261.

Hwang, Balbina Y. 2004: "Introduction: The Rise of Korea", *East Asia*, Vol. 21, No. 2, pp. 39–41.

Ikenberry, G. John and Mo, Jongryn 2013: *The Rise of Korean Leadership, Emerging Powers and International Order*, New York: Palgrave Macmillan.

Jun, Seong Ho 2014: "Monetary Authority Independence and Stability in Medieval Korea: The Koryŏ Monetary System through Four Centuries of East Asian Transformation, 918–1392", *Financial History Review*, Vol. 21, No. 3, pp. 259–280.

Kang, Tae-jun 2015: "South Korea's Quest to Be a Major Space Power", *The Diplomat*, 27 March. Available at: http://thediplomat.com/2015/03/south-koreas-quest-to-be-a-major-space-power/ [Accessed 5 November 2018].

Kim, Hyun-Wook 2011: "Obama's East Asia Policy and US-South Korea Relations", in: Melissa H. Conley Tyler and Wilhelm Hofmeister (eds.), *Going Global: Australia, Brazil, Indonesia, South Korea and South Africa in International Affairs*, Singapore: Konrad Adenauer Stiftung/Deakin, ACT: Australian Institute of International Affairs, pp. 67–74.

Kim, Samuel S. 2000a: "Korea and Globalization (Segyehwa): A Framework for Analysis", in: Samuel S. Kim (ed.), *Korea's Globalization*, Cambridge: Cambridge University Press, pp. 1–28.

Kim, Samuel S. 2000b: "Korea's Segyehwa Drive: Promise versus Performance", in: Samuel S. Kim (ed.), *Korea's Globalization*, Cambridge: Cambridge University Press, pp. 242–281.

Kim, Samuel S. 2006: *The Two Koreas and the Great Powers*, Cambridge: Cambridge University Press.

Lee, Hyunjung 2015: *Performing the Nation in Global Korea: Transnational Theatre*, London: Palgrave Macmillan.

McCourt, David M. 2011: "Role-Playing and Identity Affirmation in International Politics: Britain's Reinvasion of the Falklands, 1982", *Review of International Studies*, Vol. 37, No. 4, pp. 1599–1621.

O'Neill, Andrew 2015: "South Korea as a Middle Power: Global Ambitions and Looming Challenges", in: Scott A. Snyder (ed.), *Middle-Power Korea: Contributions to the Global Agenda*, New York: Council on Foreign Relations Press, pp. 75–89.

Oberdorfer, Don 1997: *The Two Koreas: A Contemporary History*, Reading, MA: Addison-Wesley.

Oh, Kongdan 1999: "South Korea's Foreign Policy: A Dolphin among Whales?", *Paper presented to the 14th Annual Conference of the Council on US-Korean Security Studies*, 27–30 October, Arlington, VA.

Robertson, Jeffrey 2016: *Diplomatic Style and Foreign Policy: A Case Study of South Korea*, London/New York: Routledge.

Robinson, Michael E. 2007: *Korea's Twentieth-Century Odyssey: A Short History*, Honolulu: University of Hawaii Press.

Rozman, Gilbert 2012: "South Korean National Identity: A Six-Dimensional Analysis", in: Gilbert Rozman (ed.), *East Asian National Identities: Common Roots and Chinese Exceptionalism*, Stanford: Stanford University Press, pp. 47–72.

Schattle, Hans 2014: "Global Citizenship as a National Project: The Evolution of Segye Shimin in South Korean Public Discourse", *Citizenship Studies*, Vol. 19, No. 1, pp. 53–68.

Shin, Gi-Wook and Choi, Joon Nak 2009: "Paradox or Paradigm? Making Sense of Korean Globalization", in: Chang Yun-shik, Hyun-Ho Seok and Donald L. Baker (eds.), *Korea Confronts Globalization*, London/New York: Routledge, pp. 250–272.

Snyder, Scott A. 2018: *South Korea at the Crossroads: Autonomy and Alliance in an Era of Rival Powers*, New York: Columbia University Press.

Tudor, Daniel 2014: "Korea Is a Dolphin, Not a Shrimp", *Korea Joongang Daily*, 3 February. Available at: http://mengnews.joins.com/view.aspx?aId=2984326 [Accessed 5 November 2018].

Watson, Iain 2016: "From Middle Power to Pivot Power: Korea as an Arctic Observer in the Age of Eurasia", *Pacific Focus*, Vol. 31, No. 3, pp. 333–356.

Wendt, Alexander 1999: *Social Theory of International Politics*, Cambridge: Cambridge University Press.

Zehfuss, Maja 2001: "Constructivism and Identity: A Dangerous Liaison", *European Journal of International Relations*, Vol. 7, No. 3, pp. 315–348.

2 Korea in the World

Identity and Foreign Policy

Before the dawn of the Western Westphalian international order of sovereign nation-states in East Asia in the 19th century, Korea was close to the civilizational center in the Sino-centric world as a major cultural and economic hub. Beginning with Qing China's downfall, this changed and ended in Joseon Korea's colonization by the Japanese Empire, after unsuccessful attempts at domestic modernization. During the Cold War, South Korea served as a bridgehead of the US-led so-called free world against communism, despite being ruled by military dictators until the late 1980s. Since then South Koreans have been redefining their place in the world with an unprecedented global outreach in cultural, economic, and security-related matters. But what is South Korea's place on the global stage in the eyes of others and Korea itself?

As will be shown, there are hopes from some scholars for South Korea to become a future leader of the liberal international order as a middle power or a building block for deeper regional integration. Others point to a resurgence of Korean nationalism and a desire for external status affirmation. It will become apparent in the following that there is a double gap in the research on South Korean foreign policy and international identity. On one hand, there is no comprehensive empirical assessment of international state identity or self-understanding of South Korea as an emerging global power, with potentially globalist, regionalist, and nationalist self-understandings at play. On the other hand, identity as an analytical concept has been applied in various divergent ways in the study of South Korean foreign policy but with an overall lack of taking South Korean perspectives and ultimately agency into account. This is where the conceptual approach to international state identity developed later in this book will contribute to a solution.

The Sino-Centric World

Before the Western encroachment on Northeast Asia in the 19th century, Joseon Korea had for centuries been part of the Sino-centric world order with the Chinese emperor, the Son of Heaven, at the political, economic, and cultural center. Relations between China and secondary states were organized through a tributary system based on a formal status hierarchy but allowing for significant local autonomy. Key institutions within this system were the recognition by the superior state through "investiture," which signified peaceful relations and mutual protection against external threats,[1] as well as the right of entry into China and the duty to send regular tribute missions to the Chinese court from the inferior state.[2] Early Joseon, for example, sent up to three missions a year to the Ming dynasty, not just as a diplomatic duty but also for economic exchange and cultural borrowing.[3] For this close relationship, Joseon Korea used the term *sadae*, which translates to "serving the great."[4]

Within the Sino-centric world, the source for a country's ranking was determined not so much by its material power but by its cultural similarity to the Chinese civilizational center.[5] Joseon Korea ranked first in the hierarchy during the Ming dynasty from 1368 to 1644, as it was from early on emulating, conforming to, and even trying to excel in the mastery of Chinese ideas, principles, and norms.[6] Ming–Joseon relations had been close right from the beginning. In 1388 General Yi Seong-gye, an acclaimed military leader who had fought against Japanese pirates, the Chinese Red Turban rebels, and as a military commander of Yuan China in Manchuria,[7] carried out a coup d'état and explicitly sought investiture from the Ming court to garner domestic legitimacy for his rule.[8] He subsequently became King Taejo of Joseon in 1392, establishing Korea's longest ruling dynasty.[9]

Later in the early 17th century with the rise of Jurchen or Manchu state in the northern border regions, and despite careful attempts to avoid being drawn into conflicts there, Joseon Korea had to comply with Ming requests for military support and dispatched 10,000 troops, which were advised to surrender to the mighty Manchus if necessary to avoid being drawn into larger conflict.[10] When the Ming finally fell to Manchu invaders, who, in turn, established the Qing dynasty, it still came as a shock to Joseon Korea, who perceived this as "tantamount to the disintegration of civilization."[11] Facing the center of their world ruled by "northern barbarians,"[12] Confucian Korea's "sense of cultural competition with China"[13] from the centuries before now turned into a feeling of being "culturally superior"[14] and indeed the "last bastion of Confucian orthodoxy,"[15] as well as the "sole carrier of the civilized tradition which was lost in China."[16] This notion was also reflected in the term *small China* or *sojunghwa*. Despite these reservations, the Manchu-led Qing dynasty ultimately enforced tributary relations with Joseon through invasions in 1627 and 1637,[17] after which the Qing did not further interfere with Korean affairs until the late 19th century. According to Phillips, the Qing acted toward Joseon within established rules in the beginning, because "given Korea's unimpeachable Confucian credentials, its endorsement of the Manchus as legitimate heirs to the Mandate of Heaven represented a potential propaganda coup for the ascendant Qing Dynasty."[18]

In contrast to shrimp narratives that focus on Korea's perennial victimhood at the hands of external forces, Kang classifies the Sino-centric tributary system as relatively stable and peaceful, at least among the major Sinicized powers, China, Japan, Korea, and Vietnam.[19] Apart from low-intensity skirmishes with

Japanese pirates or nomadic raiders from the north, he qualifies only two conflicts involving Korea as major wars: the devastating invasions of the Korean peninsula by Hideyoshi Toyotomi from 1592 to 1598, known as the Imjin War in Korea, and the Manchu campaigns to enforce Joseon's tributary status in the early 17th century.[20] Kang thus speaks of a "longer peace"[21] throughout the last centuries for Korea and East Asia as a whole. In addition, Shin emphasizes that despite the widespread image of being a "hermit kingdom,"[22] Joseon Korea "was an active participant in the larger East Asian cultural space."[23]

This rather stable and peaceful Sino-centric world order was shaken when European imperialist powers arrived in East Asia, putting pressure on Qing China, in particular. The Qing were already facing domestic unrest when the British Empire "opened" China with the two Opium Wars from 1839 to 1842 and from 1856 to 1860, imposing preferential trade access through the so-called unequal treaties. Similarly, in 1853, American commodore Matthew C. Perry forced the opening of protectionist Tokugawa Japan through gunboat diplomacy with his famed Black Ships. Despite a few earlier Western expeditions, Joseon Korea did not yet see the same fate and was still trying to preserve its autonomy and Confucian orthodoxy. But the decade of the 1860s was also for Koreans the beginning of their enforced modernization: "[h]ere is the beginning of modern Korea: its leaders no longer could shape events as they wished. For the first time in its history, the country was shaped from without more strongly than from within."[24] In all three major East Asian states the old regimes tried to respond to the new challengers with programs of reform in the name of "restoration,"[25] for China and Korea very much in vain but in the case of Meiji Japan's modernization ultimately successful.

Joseon's worldview and foreign policy toward the Qing, European powers, and missionaries can be illustrated with a brief event in 1865/66 at the demise of the old order. These events of 1865/66 awakened interest in Korean affairs by other actors such as the US and Japan, as it became clear that France and Russia had been increasing their influence in the region. In 1865/66, Joseon was ruled by Prince Heungseon as the regent Daewongun on behalf of his eleven-year-old son, the newly selected King Gojong. The Daewongun, having been banned from politics before because of his low-born but royal background, believed that factionalism at court was the main malady of the Joseon court and hence pushed for a dynastic revival through centralization and royal restoration.[26] Kim describes him as largely "ignorant of the outside world"[27] and assesses the Daewongun's foreign policy goals as seclusion and resistance toward Western imperialism. For the Daewongun the fate of civil war that plagued Edo Japan was a warning example, because in his eyes the Tokugawa regime had failed to resist the destructive Western influence.[28] The only relevant external power for Joseon's ruling elite remained the Qing court, to where any diplomatic matters with either France, Russia, or Japan were delegated with the request for "Chinese guidance and protection,"[29] despite Qing China's fading capabilities of meeting Joseon's expectations about dealing with the foreign intruders. Cumings summarizes this neglect for building new international ties: "[T]o those who knocked at its gates, Korea said in effect, 'We have nothing and we need nothing. Please go away.'"[30]

For the Joseon government, the spread of Christianity, in particular, had hitherto been perceived as a dangerous, alien challenge to the cultural and political hegemony of Confucianism within the country. Among the frequent and bloody persecutions of Catholic Koreans, the year 1866 stands out, when most of the French missionaries, who had only been in the country since 1834, as well as thousands of Korean Catholics were killed. The latter had allegedly reached out to the Daewongun to form a triple alliance with France and Great Britain to counterbalance Russian advances into Northeast Asia. Despite the reported initial interest of the Daewongun, the plans were never realized.[31] Not only did the conservatives at court very much disapprove of what was seen as a possible affront to Qing China, and hence a violation of the traditional East Asian order, but French missionaries, who were approached as potential intermediaries, also did not want to become involved in politics, fearing the very reprisals that were nevertheless soon to happen to them.

To give some brief context, in France Napoleon III pursued the goal of leading his country back to past glory, which translated into a project of global, imperial expansion.[32] Relying on a nationalist and Catholic support base, from 1853 to 1855 he engaged in the Crimean War with Russia over the title of protector of the Holy Land and invaded Mexico as well as China, the latter from 1858 to 1861. At the same time, in 1860, Qing China had given Imperial Russia a lease on the Liaodong peninsula, tightening already-existing border problems such as raids and uncontrolled migration in the north of the Korean peninsula, fitting traditional Korean threat perceptions. As a result of the 1866 Catholic persecutions, Imperial France sent a punitive expedition that first temporarily seized Ganghwa Island at the estuary of the Han River leading to Seoul, in a prelude to a series of gunboat interferences there that ultimately ended with the Treaty of Ganghwa Island of 1876, the "opening" of Joseon Korea by Meiji Japan.

By the end of the 19th century, the struggle over the future course of the country was at its height, with Joseon elites and people trying to recalibrate their country's place in a changing world, not without significant foreign influence from especially Japan, Russia, and China. The Donghak peasant movements of 1894, for example, fought against the Chinese-centered tributary relations of Joseon Korea, while the so-called *Gabo* reform, a modernization attempt after the example of Meiji Japan, faced opposition from anti-Japanese rebels.[33] Thirty years after Korea had been "opened," the First Sino-Japanese War from 1894 to 1895 finally buried the old East Asian world order of China-centered tributary relations. According to the terms of winning Imperial Japan, Joseon Korea was declared an independent country and thus effectively broken out of the Chinese sphere of influence. Amid this decline of China and the rise of Japan, as well as the entry of European powers, Koreans needed more than ever to redefine their position in the region and beyond.[34]

During the Russo-Japanese War from 1904 to 1905 over the influence over Manchuria and the Korean peninsula, the Joseon court remained officially neutral but fell more and more under the control of Imperial Japan.[35] After Japan had won the war against the Russian Empire, in September 1905 the Treaty of Portsmouth ensured Japanese supremacy in Korea.[36] Previously, other European powers had already recognized the Japanese claim to and increasing control over Korea in the secret Taft–Katsura Agreement between the US and Japan from July 1905 and the Anglo-Japanese Alliance from August 1905.[37] This left the Joseon court internationally sidelined and resulted in the forceful conclusion of the Protectorate or Eulsa Treaty in November 1905, which gave Japan full control over Korea's external relations.

The ensuing international isolation of Korea was demonstrated to a global audience during the so-called Hague Secret Emissary Affair (*heigeu teuksa-sageon*) in 1907. Just before the Second Hague Peace Conference in 1907, Emperor Gojong, who reportedly had never agreed to the Protectorate Treaty,[38] sent secret emissaries to the Netherlands "to expose the injustice done [to] Korea and ask for redress."[39] The delegation was refused entry on grounds of not having diplomatic sovereignty,[40] and one of the emissaries, Yi Jun, even committed suicide in the days after.[41] The whole affair generated significant international media attention but ultimately only led to the Japanese authorities forcing Gojong to abdicate.

During this period the Japanese adaptation of the Western concept of a nation as a racial category also started to bear fruit in Korea. The idea of a distinct Korean nation, *minjok*, a nationalist foundation myth with an ancient ancestor Dangun at the center,[42] as, for example, in the work of nationalist historian Sin Chae-ho,[43] and an understanding of sovereignty as a "pedagogy for imperialism and colonialism"[44] started to develop.[45] Despite the efforts of reformist thinkers such as Yu Kil-chun, who had tried to reconcile the old ways with the new, this "internationalization" (*gukjehwa*) ultimately ended in the enforced adoption of Western modernity.[46] Another influential idea of the time that gained a certain prominence throughout East Asia was that of a pan-Asian unity in a global struggle against Western imperialism. According to Shin, this thinking about "[y]ellowness, as a category transcendent to the nation, was not seen as undermining the sovereignty of the nation."[47] Ultimately, Imperial Japan utilized a similar "Western threat, Eastern solidarity" narrative in order to build a regional identity that served as justification for colonial expansion into East Asia.[48] Later in postliberation Korea, this colonial link left ideas of pan-Asianism politically tainted and thus neglected until the late 1990s.

Colonization and National Division

During colonization by Imperial Japan from 1910 to 1945, Korean nationalist liberation movements such as the March First independence movement of 1919 or the later upcoming communist or radical nationalists further developed Korean nationalism, similar to the situation in other colonized countries. According to Rozman and Kim, the nation or *minjok* became the "core identity, making it the means for national survival and [. . .] a source of legitimacy."[49]

The forced assimilation by Japan made Korean nationalists more aware of distinct cultural attributes as the only way to ensure their nation's survival in the face of threatening Japanese absorption. Others, such as Korean proponents of pan-Asianism, subscribed to the Japanese theory of common Japanese and Korean ancestry, *nissen dōsoron.* According to this official colonial ideology, Japanese and Koreans descended from the same race but with the Koreans in need of Japanese guidance to achieve "civilization" and "enlightenment." Shin writes that collaborative Korean merchants and capitalists especially tapped into this narrative in a pragmatic manner, as Korea was located strategically in the center of the Japanese colonial trade network.[50]

After liberation from Japanese colonial rule, the contestation over the meaning of the Korean nation further deepened due to the establishment of two separate Korean states. In both states' official national imagination, Korea remained a perennial victim that had suffered a variety of invasions by external powers from the region.[51] In the DPRK or North Korea, Kim Il Sung's *juche* ideology, usually translated as self-reliance or self-autonomy, heavily relied on anti-imperialism and anti-Japanism. For some scholars, North Korea today can barely be seen as a communist country but more as a fascist state that advocates the preservation and protection of the racial and hence moral purity and uniqueness of the Korean race.[52] In South Korea, military strongman Park Chung Hee came to power through a coup d'état and envisioned the "modernization of the fatherland"[53] through instruments of repression as well as selective use of Korean historiography.

Central to Park's promoted nationalism was the understanding that Japan's colonization of Korea was only possible because Joseon Korea and its successor state, the Korean Empire (1897–1910) that was already heavily influenced by Japan, had been internally and externally weak. The entirely outdated Korean state thus was easy prey for the successfully modernized Meiji Japan (1868–1912). According to this perspective, colonial history proved the lesson that, for the survival of the nation, Koreans have to modernize their state after the example of the West and Japan, in particular. But Park's nationalist ideology incorporated ethnic as well as political dimensions: North Koreans, for example, were seen as brethren, as descendants of the same 4,000-year-old race, sharing the same history, language, and destiny.[54] In terms of a political nation, North Koreans did not belong to the same community and thus should get help in removing the un-Korean and alien communist rule—the only legitimate national identity was incorporated by the southern state, the ROK. Hence, there was an overarching ethnic national unity that had to be restored politically by a "clear-cut nation saving policy of independence, self-reliance and self-defense."[55]

The South Korean democracy movements of the 1980s, too, were founded on a nationalist historiography.[56] This strain of the so-called people-oriented (*minjung*) nationalism criticized Park's regime for enslaving the South Korean people and, due to its anticommunism, fostering the division of the Korean peninsula. According to this narrative, the main tasks for the nation are democratization and unification: for the democratic opposition, Park's authoritarian regime was largely a product of US containment strategies toward communism in East Asia and democratization. Correspondingly, liberating the nation meant overcoming not only Park's dictatorship but also the influence of yet another external and almost neo-imperial power, the US.[57] The key difference here is that, according to the earlier state-led nationalism, the Korean identity lies in a sacrifice for the modernization of the fatherland and for anticommunism, whereas for oppositional thinkers such behavior is seen as anti-Korean, construing the suppressed people and fighters for national self-determination as the true bearers of the nation.

Common to both self-understandings, but even for North Korean *juche* ideology, was the centrality of a national struggle for liberation from foreign dominance, be it from Japanese imperialism, Western communist thought, or perceived American neo-imperialism. This is illustrated in a quote from President Roh Moo-hyun, a former pro-democracy activist and civil rights lawyer: "No other interests or values could compete [with] those of liberating the motherland from foreign domination and securing political sovereignty."[58] Moreover, generations of Koreans until today have been raised and educated with specific ideas about the uniqueness of the Korean people, culture, and the "sanctity of [the] territory, creating a nationalist consensus among Koreans of all ages."[59] Hahm and Kim speak of two main pillars of modern South Korean self-identity: anticommunism and anticolonialism—Japan, in particular, is seen as the other for the Korean self, but anticommunism would have loomed larger during the Cold War standoff on the peninsula.[60] For them, finally, with the end of the Cold War, the balance changed and a nationalism-centered discourse resurfaced. Furthermore, Kim sees this ethnic Korean nationalism as the "perspective [that] is fixated on the artificial division of the Korean people by foreign powers, tempered only by the belief that their destiny lies in the emancipation of their political divisions through their common essence."[61]

As mentioned earlier, South Korea was usually regarded as a shrimp amongst whales, as a minor, squeezed-in player that was all too often a victim of the battles between larger forces. Due to this geopolitical situation, postliberation Korea was also attributed only limited leeway in its foreign policy choices and thus had to constantly maneuver between the more powerful regional actors.[62] The ROK's economic and security position has thus been crucially dependent on the asymmetrical alliance with the US. Both states had signed the "Mutual Defense Treaty Between the United States and the Republic of Korea" after the Korean War in 1953—and the US still holds wartime operational command (OPCON) over South Korean forces today. Son therefore attributes South Korea—what he calls the international security identities of "garrison state" and "blood-forged ally"—when describing South Korea as a foreign policy actor vis-à-vis the US.[63] When South Korean troops were sent to fight in the Vietnam War from 1964 to 1973, it was by direct request from Washington; Seoul saw the contribution mostly as a tool for alliance management[64] or for generating revenue through inclusion in the war economy.[65] Furthermore, a key aspect of Seoul's Cold War foreign policy was the quest for domestic and international legitimacy in relation to North Korea. Domestically, in the absence of democracy, authoritarian rulers opted for nationalist and anticommunist narratives to sustain their regime,[66] while internationally Pyeongyang and Seoul competed in a race for external recognition and international support as *the* Korean state.[67] This mostly ended in 1991 when both Korean states finally became members of the United Nations. In 1996 South Korea even became a member of the OECD, while North Korea was internationally isolated over its announced withdrawal from the Nuclear Proliferation Treaty (NPT) and was suffering from large-scale famines resulting in the death of over a million people.[68] In the eyes of the world, the race for international legitimacy and recognition between the two Koreas had been largely decided by now.

The Post-Cold War Period

In the late 1980s much had changed within South Korea and in the wider region. Internally, the country underwent an impressive transformation from authoritarianism to democracy, and internationally, for the first time, diplomatic relations were reestablished with China and Russia as a result of President Roh Tae-woo's *nordpolitik* initiative. Moreover, President Kim Young-sam had just started his ultimately economy-focused *segyehwa* (globalization) process, attempting to open Korea more to the outside world, when the Asian financial crisis hit South Korea and the IMF intervened. In consequence, the proud developmental narrative about the miracle from the Han River suffered substantially.

In terms of Korean international identity, however, in the literature this outreach to the world of the 1990s is not seen as a wholehearted embrace of globalization. Alford, for example, stresses the perceived threat of globalization to the social fabric of Korean societal relations.[69] Moreover, because of 1990s democratization processes, state-led and popular nationalism became disconnected, and Seoul lost its influence over public discourse on national identity.[70] Hundt and Bleiker add that in newly democratized South Korea, for the first time, "long-held and long-repressed feelings of resentment [were able] to emerge."[71] Anti-Japanese and anti-American sentiments made their way into the media, whereas previously such voices and political demands had been mostly suppressed by the military dictatorships. This was seen as an indication that the opening of the country in the 1990s was reinforcing nationalism, the "paradox of Korean globalization."[72] Whether indeed a paradox or not, Shin makes the point that the instrumentalist appropriation of globalization by the Korean state was "a familiar paradigm in Korea's history of social change"[73] and "a necessary means of enhancing national interests."[74] Saxer likewise argues that "for close to two decades Korean governments have [. . .] emphasized the aspirations of taking advantage of globalization to improve the status and international position of the country."[75]

For the area of national security, Cha speaks of "a tension between globalization pressures on the one hand and localization imperatives on the other."[76]

For Cha these structural constraints necessarily led to Korea being "firmly entrenched in a classic realist mindset [. . .] the residue of history and certain immutable geopolitical traits." Further, he describes the resulting "Korean identity [. . . as] strongly negative in strain" and "reactive nationalism"[77] fed by antiforeign Confucian values.[78] Besides, Shin notes the resurgence of another Asianism narrative through conservative intellectuals for the first time after the colonial era, that of Asian values as a defense to maintain "society and identity against the state-led globalization that promoted neo-liberal 'Western' values."[79] The Singaporean president Lee Kuan Yew had spearheaded this international debate by arguing that "what he termed Asian values such as respect for authority and consideration of the community's interests above the individual's interests could allow for economic growth and promote social harmony while avoiding the social pathologies prevalent in the West."[80] The prominent South Korean opposition figure and democracy activist Kim Dae-jung, however, argues that Lee's take on Asian culture was a selective mischaracterization based on a "myth,"[81] as "Asia has a rich heritage of democracy-oriented philosophies and traditions."[82]

Only a few years later, in 1998 the same Kim Dae-jung became the first progressive president in the country's history. Korea's relationships with two significant others, North Korea and Japan, were now recalibrated, in what was understood as a "shift in identity."[83] The sunshine-policy engagement of Pyeongyang as well as the steps toward reconciliation with the former colonizing power, Japan, were very much set to change the rules of the regional power game, although neither had a lasting effect beyond Kim's presidency (1998–2003). Conservatives, in particular, remained skeptical of Kim's advances with North Korea, due to their fear of a subsequently weakened US–ROK alliance. According to Snyder and Glosserman, "they saw the alliance as a necessary support for South Korea's diplomatic efforts in the region and focused more on raising South Korea's profile on the global stage rather than diplomatic initiatives to strengthen regional cooperation in East Asia."[84]

In contrast, President Roh Moo-hyun (2003–2008), also from the progressive camp like his predecessor, advocated a focus on regional integration and envisioned a balancer (*gyunhyeongja*) role for Korea, very much equidistance from both the US and China, which was initially very popular with the South Korean public[85] but with little external recognition from Japan and China and effectively leading to a weakening of ties with Washington.[86] As a reaction, the conservative Park Geun-hye, the daughter of former authoritarian leader Park Chung Hee and later South Korean president herself, accused President Roh of "sabotaging Korean national identity."[87] In addition, as Ryu elaborates, President Roh pursued a strategy aiming at more independence or autonomy (*jaju*) from the US, in particular with regard to Korea's own military capabilities and the transfer of mission control.[88]

Ryu further argues that the emphasis on a more independent foreign policy, especially vis-à-vis the US was typical of the generation of democratization and student protest that came to power during the progressive presidencies.[89]

This is the so-called 386 generation of those born in the 1960s, who went to university in the 1980s and were 30 years old during the 1990s, when the term was first coined. President Roh's ROK National Security Council had also declared that Korea is nowadays "a major actor, not a subordinate variable" in Asia.[90] According to Cho and Park, it has been a key feature of South Korean and Japanese foreign policies not to be "content to play the role of followers"[91] of neither the US nor China and therefore to propose "alternative region-making initiatives appealing to domestically contested views on how best to seek autonomy from the region's Great Powers as a way to enhance their political standing domestically and regionally."[92] They conclude that Japan and South Korea are both pursuing a vision of more regional autonomy outside of US-led hub-and-spoke system of alliances in East Asia.

With the two progressive presidencies, the first in South Korean history, a new form of progressive Asianism came to the fore, but this time it was based on a confidence in the economic rise of East Asia and a changing perception of North Korea from a threat to an object of pity.[93] This is reflected in Campbell's long-term study about young South Koreans, for whom the ethnic dimension of Korean national identity lost its relevance.[94] Given the potential costs of reunification with the impoverished North, many South Koreans also became highly skeptical of any short- or medium-term solution to the division of the Korean Peninsula. This is also the reason why young South Koreans, who are used to economic prosperity and a relatively stable security environment, have been described as more security conservative than previous generations, putting them in the same group with the strongly pro-American, anticommunist 60-plus generation.[95]

Campbell, moreover, observes ROK-centered, oppositional motivations and protest, diagnosing a change in national interests and has thus indirectly raised the question about the underlying change of identity formation. Environmentalism, personal well-being, historical issues such as the disputes about the naming of the East Sea/Sea of Japan, and, most prominently, the Dokdo/Takeshima territorial dispute are on the center stage nowadays. Furthermore, Campbell alludes to the still latent anti-Americanism as expressed by the protests against the import of US beef in early 2008 and to the high appeal of status-related goals such as improving South Korea's national brand, global ranking, or leadership capabilities in the international community. She concludes that "[f]or many young people, pride in ROK's achievement and its growing international importance is a key part of their national identity, and as such they demand appropriate recognition for South Korea and respect from the international community."[96]

A major development in this regard was also the unprecedented global circulation of Korean popular culture, known as *hallyu* or the "Korean Wave."[97] Starting in the 1990s South Korea successfully developed transnationally successful music, television, and film industries, systematically supported by the government in an effort to find new products and export markets. Globalization had initially been seen as a dangerous "Western cultural invasion [. . .]

amplified by the uncertainty of the competitiveness of Korean popular culture,"[98] and until 1998 even Japanese cultural imports were banned in South Korea. The success of the Korean Wave changed this and became an international testament to the risen status of Korean culture. As a consequence, at least for younger generations, "globalisation [became] a tool for building South Korea's national brand and to gain international leadership."[99]

Likewise, with the help of public opinion data, Denney and Friedhoff show how for Koreans the time of being a shrimp among whales is coming to an end, with their country "assuming its rightful place in the hierarchy of nations."[100] According to their findings, there is the rise of a new nationalism as a "manifestation of a new national identity and a natural outgrowth of the country's material development and newfound confidence."[101] Many Koreans now see their country overcoming "the stigma of being a victim of colonization"[102] and support efforts to raise Korea's international visibility and recognition as an important state among the first ranks of world affairs. Kim, for example, emphasizes that many South Koreans focus on their country's "impressive achievements, from outperforming many of the world's most advanced nations in fields as varied as technical innovation and hosting the Olympics."[103] Rozman also points toward a more and more confident and status-driven South Korean self-understanding on the world stage:

> South Koreans aspire to a proud national identity based on something more than economic accomplishment, democratic rights, and a well-respected place in international society. Success on those dimensions has not quenched the thirst for [a] deeper appreciation of the South Korean state and for the Korean people for their special significance in the past, present, and future of Northeast Asia.[104]

On the other hand, Ducke suggests that precisely because of the strong international-status inclination of Korea, Seoul should be more compliant with expectations from the international community.[105] What the aforementioned surveys do not reveal, however, is whether and how these attitudes relate to concrete foreign policy decisions and the wider foreign political discourse.

In any case, the most notable recent period in terms of global outreach was the presidency of the conservative Lee Myung-bak from 2008 to 2013. In 2009 the newly inaugurated president presented his national security strategy called *Towards a Global Korea*, aimed at performing middle-power diplomacy "not only to improve Korea's image, but also to exert influence on the international system."[106] The national vision drew the picture of a globally interconnected, modern, and responsible member of the international community that is prone to assume leadership for the first time in such traditional OECD areas as international peacekeeping operations (PKOs), official development assistance (ODA), and climate diplomacy.[107] A fundamental image evoked was that of a model student that succeeded in fast modernization and development, and is hence in a position to finally give back. Moreover, there was an ambition to serve as a bridge between the developed world and developing countries. This very assertive idea about Korea seems to mirror Campbell's assessment of younger Koreans' proud focus on South Korea's more recent developmental achievements. Rozman and Kim moreover see the appeal of traditional self-understandings undermined by Lee's election and the rise of an identity focused on shared values such as democracy or free-market capitalism.[108] The "Global Korea" image is seen as a more positively defined self-image, quite in contrast to the victimization inherent in more traditional understandings. According to Cho and Park, President Lee's foreign policy and national security advisor expressed that "the Lee government's goal was to enhance South Korean *gukgyuk* [*sic*] (or national prestige) on the global scene."[109]

Under Lee's presidency, Seoul did successfully bid for its very own approach to sustainable development by establishing the Global Green Growth Institute (GGGI), and hosted multiple multilateral summits such as the G20 in 2010, the Fourth High Level Forum on Aid Effectiveness (HLF-4) in 2011, and the Nuclear Security Summit (NSS) in 2012. Last, Korea also joined a newly formed platform for middle-power cooperation, MIKTA, which stands for Mexico, Indonesia, South Korea, Turkey, and Australia.[110] South Korea is also the first country to have ever developed from an aid recipient to an international aid donor. In contrast to the shrimp narratives, Korea's newfound status was thus described with the metaphor of a "dolphin."[111] With the inauguration of President Lee, Chun argues that a new "globalist school of thought" emerged in South Korean foreign policy making, but he also concedes that all presidential candidates had voiced similar visions about Korea's expanding global horizon.[112] Chun further suggests that globalists in Seoul favored such roles as "mediator, facilitator of co-operation, convener in international co-operation, agenda-setter, bridge-builder, or switcher in international networks."[113] On the other hand, he sees the main challenge for a globally oriented South Korea in a lack of globalist identity, as "the South Korean public still adheres to its old self-perception to confine South Korea's role and target regions."[114]

To put this into some context, in a 2010 national identity poll[115] more than 75 percent of South Korean respondents indeed viewed their country as a middle power that, according to half of all respondents, should serve as a mediating bridge between the developed and developing worlds, but significantly fewer saw their country already as a "leader."[116] In the corresponding larger long-term survey on the Korean public and national identity, which was undertaken by the East Asia Institute (EAI) 76.8 percent in 2010 and 69.4 percent of respondents identified their country also as a middle-ranking country (*jungjinguk*), and 19.9 percent in 2010 and 27.2 percent in 2015 perceived their country still as a small nation (*yaksoguk*).[117] Furthermore, about half of the respondents (40.7 percent in 2005, 49.4 percent in 2010, and 50.9 percent in 2015) agreed or strongly agreed with the statement "Our country should increase aid to poor countries."[118] In other findings, a strong majority of the respondents (67 percent in 2005, 50.5 percent in 2010, and 65.1 percent in 2015) agreed or strongly agreed with the statement that "Korea is not being properly treated on the international stage."[119] With regard to regionalism or Asianist positions, 54 percent in 2010 and 68.2 percent in 2015 supported the formation of a regional, borderless union among China, South Korea, and Japan similar to the European Union,[120] with two-thirds of respondents generally in favor of such a regional integration even without the involvement of the US.[121]

In terms of a globalist self-conception there is, however, a significant trend of increasing support for compliance with the decisions and rules of international organizations: 37.5 percent of respondents in 2005, 51.4 percent in 2010, and 59.1 percent in 2015 agreed or strongly agreed with this position.[122] A similar survey of South Korean 200 policy makers' attitudes toward ODA, undertaken by the Asan Institute in 2016, found that, while 99.5 percent of elites strongly or moderately supported South Korean ODA, there was somewhat less support (86.5 percent)[123] from the Korean public.[124] There were also minor differences between progressives and conservatives when it came to the reasons behind their support for ODA. While 46.8 percent of progressives gave the contribution to international peace and stability as their main reason, 38.7 percent of conservatives named the improvement of South Korea's image and foreign relations as their main motivation.[125]

The earlier mentioned EAI, a well-established think tank with close links especially to the Ministry of Foreign Affairs, was one of the main drivers behind the recommendation for the government to consider middle-power concepts for Seoul's future diplomatic trajectory,[126] and so was South Korea's main ally, the US. Lee Sook-jong, then the EAI's president, argued that

> South Korea's growing multilateral diplomacy has been pursued at no cost to its alliance ties with the United States. On the contrary, South Korea has utilized its political and diplomatic ties with the United States in promoting its regional and global influence.[127]

Accordingly, President Lee Myung-bak also shifted the administration's diplomatic focus back to the US–ROK alliance and pursued a rather hardline stance toward North Korea, effectively ending the era of engagement with the north supported by his two predecessors.

After all, there remains an ambivalent picture about the emergence of a new post-Westphalian globalist identity that is transcending national boundaries, especially if one remembers President Roh's focus on a more independent and regionalist foreign policy as well as the earlier assessments of the paradox of Korean globalization, where the global order was mostly seen as a competition between nations.[128] Kalinowksi and Cho, for example, argue that Korea's global strategy with regard to ODA and international agenda setting has been dominated by national economic interests to safeguard the competitiveness of Korean businesses:

> Korea's global strategy is dominated primarily by short-term economic interests of Korean businesses. This pattern can be explained by the path

> dependency of Korea's mercantilist development strategy. Ironically, the institutional legacy of Korea's global economic success has become an obstacle to a broader global political role that is recognized and accepted by countries around the world.[129]

Even Ikenberry and Mo, who suggest South Korea's global leadership as the future of the international liberal order,[130] concede that an international profile and extensive multilateralism is "not an obvious outcome"[131] for a nation formerly known as the hermit kingdom that has suffered various foreign invasions throughout history and was very recently a war-torn country. The tensions between the different identity conceptions are especially visible in the relationship with Japan. Here, the postcolonial objections against external forces and the significance of self-autonomy collide with the capitalist and globalist self-understanding that President Lee tried to address when he, for example, suggested pragmatic and future-oriented cooperation with Japan. But even this trait can have conflictual consequences insofar as more self-assertive self-understandings may pave the way for status battles, for example, in the case of the naming of the East Sea/Japan or Dokdo/Takeshima: "Sensing a lack of international profile and recognition, Korea quite literally wants to be put on the map by demanding the international classification of Korean names for geographic features."[132]

Today it is widely agreed that South Korea is a middle power in regional and global affairs. In the early 2000s, Seoul's global role had already been categorized as an emerging middle power,[133] an actor that is still unlikely to lead and thus more likely to be led.[134] Since at least 2008, however, when Seoul successfully became a member of the G20, most scholars[135] agreed with Nye that South Korea had become an "important middle-ranking power in global affairs"[136] and a more confident emerging global power with its own leadership ambitions.[137] The definition of what constitutes a middle power is contested in the literature,[138] but, as neither great nor small powers, these countries are supposed to have the will and capacity to a pursue niche diplomacy that targets limited policy fields or an instrumental interest in multilateral cooperation and international institutions for constraining more powerful states.[139] Therefore, they are said to have some rule-making power, which makes them more than mere rule takers, even though this can be said for small powers, such as Qatar, Norway, or New Zealand, as well[140]. Still, South Korea today is regarded as an exemplary middle power and a regional hub for trade and innovation with a strategic position in the global economy. Cooper even sees South Korea as a "stand out case" of rising middle powers, "not only as a specific non-traditional middle power but also as a leader in a wider East Asia 'wave.'"[141] Rozman, too, emphasizes Korea's "special opportunity to play a critical role in the emergence of substantial regional linkages"[142] precisely because of its geographical regional centrality. In contrast, according to Kim's survey of foreign policy experts, Seoul's middle powermanship is underpinned by a "self-perception as a newly advanced economy and mid-ranked global power, capable of making a distinctive contribution to the global common good,"[143] while the experts were quite wary about the concept of middle power in South Korean diplomacy overall. Kim thus concludes that "there has been much confusion and analytical ambiguity over South Korea's articulation of a middle-power posture."[144]

In addition, there appear to be very different uses of the concept of identity with regard to South Korea as a foreign policy actor, as well as a widespread but merely anecdotal use of role-theoretical conceptual language. Son, for example, without further definition or explanation of the concept, introduces "international security identities" such as "garrison state" or a "blood-forged ally" when describing South Korea as a foreign policy actor vis-à-vis the US.[145] It will become apparent in the following discussion that what Son refers to is better described as "international roles" or "international role identities." Easley's comparative analysis of South Korean and Vietnamese national identities in light of their middle-power foreign policies[146] simply looked at similarities between different national identities and infers from assumed closeness or distance that this also creates a specific "geopolitical distance."[147] But the case at hand of the Korean Peninsula shows that such simplified models are of little explanatory value: There are two Korean states that share the same people, language, and history, but how does that explain any of both Koreas' foreign policies? As already illustrated, both Korean states actually claim to provide the better nation-state platform for the Korean nation.[148] Last, Easley's understanding of identity as national self-conceptions of domestic elites remains rather vague and anecdotal, although he himself concedes that it is very difficult to track identity "accurately and reliably."[149] For Kim, who has looked primarily at Seoul's developmental policy, middle power is a "diplomatic identity" and a "policy frame," as well as a "self-identity"[150] that remained too vague and confusing for Korea's regional situation, aspirations, and strategic imperatives. Hence, Kim argues for the adoption of different concepts, such as a "creative and constructive power," which she understands as "an aspiration for the country to play a 'creative' role and exert 'constructive' influence to lessen security tensions in the region."[151] While there is much to agree with Kim's assessment of the vague and inconsistent use of middle power in the political and academic discourse, the confusion seems to mainly stem from a basic lack of conceptualization of what identity and role mean in the context of nation states and foreign policy.

Taking up the challenge of providing the field with comprehensive analytical tools to study identity, Rozman presents a six-dimensional approach to investigate national identity comparatively in East Asia.[152] He introduces several descriptors such as ideological (left, right, centrist), temporal (prewar, postwar, post-Cold War) or sectoral (culture, economics), which reveal his treatment of identity as a mere set of properties, leaving the constitutive and formative aspects of identity completely aside. The closest of Rozman's categories to international identity is his horizontal dimension, which focuses on the respective relations of Seoul with the US, Northeast Asia, and the whole world. After all, Rozman's approach is only a descriptive and external categorization, without any theorization of his understanding of identity. In contrast, Kim, aware of the need for an approach bridging levels of analysis, calls for a redefinition of national identity along the lines of what he calls synthetic national identity theory, because most IR approaches "demand [. . .] either domestic level factors or the international system as the determinant in framing foreign policy."[153] Even though his attempt to bridge levels of analysis has to be welcomed, his suggested "synthetic theory" falls short of conceptual clarity and richness, leaving open what and how to synthesize.

In conclusion, we arrive at a double gap in the literature. The first gap is empirical about the content of South Korean state identity as a rising power. As yet, there is no comprehensive assessment of the international state identity of South Korea as an emerging global power. Korea's place in the world has developed, but for most of the time Korea was not simply a small and marginal shrimp and neither was its self-understanding. Before the dawn of the Western Westphalian international order of sovereign nation-states, Korea was close to the civilizational center in the Sino-centric world as a major cultural and economic hub. Beginning with the Qing China downfall, this changed and ended in Joseon Korea's colonization by the Japanese Empire, after unsuccessful attempts at domestic modernization. During the Cold War, South Korea was, for the most part, ruled by military dictators, and the country served as a bridgehead of the US-led so-called "free world" against communism. Since the 1990s South Koreans have been redefining their place in the world with an unprecedented global outreach in cultural, economic, and security-related matters. In the early 21st century, South Korea is now showing a newfound international confidence and has voiced global and regional leadership ambitions. When it comes to the related literature, there is an agreement that the proverbial shrimp is no more, but it is unclear what has changed in South Korean international state identity and how. Given the above literature review, it seems still quite possible that South Korea's global outreach is underpinned by regionalist, nationalist, postcolonial, and globalist self-understandings. Hence, the question about South Korea's international state identity in the context of its global foreign policy remains an empirical question deserving further investigation. The potential influence of globalist, postcolonial, and pan-Asian identity narratives thus has to be of concern as the three subquestions for this study.

The second gap is conceptual and consists of a lack of agency-oriented approaches to studying the state identity of South Korea as an international actor. Identity as an analytical concept in IR has been applied in various divergent ways in the study of South Korean foreign policy, often conflated with the concept of role, but with an overall lack of taking South Korean perspectives and ultimately agency into account. Moreover, the often-referred-to structural notion in assessments of South Korean status seems rather overvalued. Not every small state is automatically weak,[154] and the historical vulnerability in the Northeast Asian geopolitical setting may even have served as an opportunity for Seoul as well. Woo, for example, has shown how South Korean governments have had "remarkable leverage on the big ally"[155] precisely because the "United States has historically been willing to pump financial resources into the country,"[156] enabling Seoul to capitalize on Washington's strategic and security concerns in Northeast Asia. Moreover, the challenging regional geopolitics have been a constraining reality for any state in the region, for neither rising China, the global superpower US, nor maverick North Korea can easily project their power there.[157]

It is the purpose of this book to close this double gap by providing a conceptually sound and empirically rich agency-oriented inquiry into the international state identity of Korea as an emerging global power. In order to do so, identity as an analytical concept in IR is discussed next, after which the role-theoretically inspired approach focusing on self-identifications and particular self-images in South Korea's global foreign policy are developed.

Notes

1. Kim Haboush, JaHyun 2001: *The Confucian Kingship in Korea*, New York: Columbia University Press, p. 21.
2. Kang, David C. 2010: *East Asia before the West: Five Centuries of Trade and Tribute*, New York: Columbia University Press, p. 56.
3. Lee, Ki-baek 1984: *A New History of Korea*, Seoul: Ilchokak Publishers, p. 189.
4. Ibid.
5. Kang, D.C. 2010: p. 54.
6. Kim Haboush, J.H. 2001: p. 22.
7. Lee, K.B. 1984: p. 165.
8. Ibid.: p. 189.
9. Cumings, Bruce 2005: *Korea's Place in the Sun: A Modern History*, New York: W. W. Norton Company, p. 44.
10. Lee, K.B. 1984: p. 215.
11. Kim Haboush, J.H. 2001: p. 23.
12. Cumings, B. 2005: p. 78.
13. Kim Haboush, J.H. 2001: p. 23.
14. Lee, K.B. 1984: p. 216.
15. Kim Haboush, J.H. 2001: p. 23.
16. Ibid.: p. 24.
17. Cumings, B. 2005: p. 78.
18. Phillips, Andrew 2018: "Contesting the Confucian Peace: Civilization, Barbarism and International Hierarchy in East Asia", *European Journal of International Relations*, Vol. 24, No. 4, p. 755.
19. Kang, D.C. 2010: pp. 82ff.
20. See also Kang, D.C. 2010: p. 83.
21. Ibid.: p. 82.
22. Griffis, William Elliot 1882: *Corea: The Hermit Nation*, New York: Scribner.
23. Shin, Gi-Wook 2005: "Asianism and Korea's Politics of Identity", *Inter-Asia Cultural Studies*, Vol. 6, No. 4, p. 617.
24. Cumings, B. 2005: p. 86.
25. Ibid.: pp. 86–87.
26. Kim, Key-hiuk 1980: *The Last Phase of the East Asian World Order: Korea, Japan, and the Chinese Empire, 1860–1882*, Berkeley: University of California Press, p. 44.
27. Ibid.

28. Ibid.
29. Ibid.: p. 48.
30. Cumings, B. 2005: p. 87.
31. Ibid.
32. See also Kim, Yong-koo 2001: *The Five Years' Crisis, 1866–1871: Korea in the Maelstrom of Western Imperialism*, Incheon: Circle Books, p. 24.
33. Mishra, Sandip Kumar 2011: "Korean National Identity Discourse: Evolution, Colonialism, and Division", *The Journal of Northeast Asian History*, Vol. 8, No. 2, p. 17; Shin, Gi-Wook 2008: "Nation, History, and Politics", in: Hyung Il Pai and Timothy R. Tangherlini (eds.), *Nationalism and the Construction of Korean Identity*, Berkeley: Institute of East Asian Studies, University of California, p. 149.
34. Shin, G.W. 2005: p. 618.
35. See also Schmid, Andre 2002: *Korea between Empires, 1895–1919*, New York: Columbia University Press, p. 28.
36. Eckert, Carter J., Lee, Ki-baik, Lew, Ick, Young, Robinson, Michael and Wagner, Edward W. 1990: *Korea Old and New: A History*, Seoul: Ilchokak Publishers, p. 239.
37. Ibid.: p. 238.
38. Eckert, C.J. et al. 1990: p. 240.
39. Ibid.
40. Lee, K.B. 1984: p. 311.
41. Cumings, B. 2005: p. 145.
42. Rozman, Gilbert and Kim, Andrew 2012: "Korean National Identity: Evolutionary Stages and Diplomatic Challenges", in: Gilbert Rozman (ed.), *East Asian National Identities. Common Roots and Chinese Exceptionalism*, Stanford: Stanford University Press, p. 202f; Moon, Chung-in and Suh, Seung-won 2005: "Security, Economy, and Identity Politics between Japan and South Korea under the Kim Dae-jung Government", *Korea Observer*, Vol. 36, No. 4, pp. 561–602; Em, Henry H. 2013: *The Great Enterprise: Sovereignty and Historiography in Modern Korea*, Durham, NC: Duke University Press, p. 77.
43. Robinson, Michael E. 1984: "National Identity and the Thought of Sin Ch'aeho: Sadaejuŭi and Chuch'e in History and Politics", *Journal of Korean Studies*, Vol. 5, pp. 121–142.
44. Em, H.H. 2013: p. 3.
45. Kim, Hun Joon 2011: "Political Legitimacy in South Korea", in: John Kane, Hui-Chieh Loy and Haig Patapan (eds.), *Political Legitimacy in Asia: New leadership Challenges*, New York: Palgrave Macmillan, p. 227.
46. Ha, Young-Sun 1999: "The Historical Development of Korean Globalization: Kukchehwa and Segyehwa", in: Chung-in Moon and Jongryn Mo (eds.), *Democratization and Globalization in Korea: Assessments and Prospects*, Seoul: Yonsei University Press, pp. 161–162.
47. Shin, G.W. 2005: pp. 619–620.
48. Ibid.
49. Rozman, G. and Kim, A. 2012: p. 203.
50. Shin, G.W. 2005: p. 620.
51. Rozman, Gilbert 2012: "South Korean National Identity: A Six-Dimensional Analysis", in: Gilbert Rozman (ed.), *East Asian National Identities: Common Roots and Chinese Exceptionalism*, Stanford: Stanford University Press, p. 54.
52. See also Myers, Brian R. 2010: *The Cleanest Race: How North Koreans See Themselves and Why It Matters*, Plymouth: Melville House Publishing; Mishra, S.K. 2011: p. 23.
53. Shin, G.W. 2008: p. 152.
54. Mishra, S.K. 2011: pp. 24–25; Shin, G.W. 2008: p. 152.

55. Cited in Shin, G.W. 2008: p. 154.
56. Rozman, G. 2012.
57. Shin, G.W. 2008: p. 161.
58. Moon, Chung-in and Li, Chun-fu 2010: "Reactive Nationalism and South Korea's Foreign Policy on China and Japan: A Comparative Analysis", *Pacific Focus*, Vol. 25, No. 3, p. 334.
59. Hermanns, Heike 2006: "The Korean Nationalist Discourse Online: An Example of Public Diplomacy in the Internet Era?", *Conference Paper of the 16th Biennial Conference of the Asian Studies Association of Australia at the University of Wollongong*, 26–29 June, p. 7.
60. Hahm, Chaibong and Kim, Seog-gun 2006: "Remembering Japan and North Korea: Politics of Memory in South Korea", in: Gerrit W. Gong (ed.), *Memory and History in East and Southeast Asia: Issues of Identity in International Relations*, Washington, DC: CSIS, p. 101.
61. Kim, Andrew 2011: "South Korean National Identity and Its Strategic Preferences", *Joint U.S.-Korea Academic Studies: Korea's Domestic Policies and Their Influence on Asia, Emerging Voices*, Vol. 22, p. 10. Available at: www.keia.org/sites/default/files/publications/emergingvoices_final_andrewkim.pdf [Accessed 5 November 2018].
62. De Ceuster, Koen 2005: "Pride and Prejudice in South Korea's Foreign Policy", *Copenhagen Journal of Asian Studies*, Vol. 21, pp. 64–90; Kang, Sung-Hack 2011: *Korea's Foreign Policy Dilemmas: Defining State Security and the Goal of National Unification*, Kent: Global Oriental.
63. Son, Key-young 2011: "From a Garrison State to a Humanitarian Power? Security Identities, Constitutive Norms and South Korea's Overseas Troop Dispatches", *The Korean Journal of Defense Analysis*, Vol. 23, No. 4, pp. 557–573.
64. Hwang, Balbina Y. 2012: "Korea and PKO: Is Korea Contributing to Global Peace?", in: Scott A. Snyder (ed.), *Global Korea: South Korea's Contributions to International Security*, New York/Washington, DC: Council on Foreign Relations, p. 15.
65. Breuker, Remco E. 2009: "Korea's Forgotten War: Appropriating and Subverting the Vietnam War in Korean Popular Imaginings", *Korean Histories*, Vol. 1, No. 1, pp. 40–41.
66. Kim, H.J. 2011: pp. 220ff.
67. Koh, Byung Chul 2001: "The Foreign and Unification Policies of the Republic of Korea", in: Kil Soong Hoom and Moon Chung-in (eds.), *Understanding Korean Politics: An Introduction*, Albany: State University of New York Press, p. 232.
68. See also Noland, Marcus, Robinson, Sherman and Wang, Tao 2001: "Famine in North Korea: Causes and Cures", *Economic Development and Cultural Change*, Vol. 49, No. 4, p. 741.
69. Alford, C. Fred 1999: *Think No Evil: Korean Values in the Age of Globalization*, Ithaca: Cornell University Press.
70. He, Baogang 1999: "The Role of Civil Society Defining the Boundary of a Political Community: The Cases of South Korea and Taiwan", *Asian Studies Review*, Vol. 23, No. 1, pp. 27–48.
71. Hundt, David and Bleiker, Roland 2007: "Reconciling Colonial Memories in Korea and Japan", *Asian Perspective*, Vol. 31, No. 1, p. 68.
72. Ibid.
73. Shin, Gi-Wook 2006: *Ethnic Nationalism in Korea: Genealogy, Politics, and Legacy*, Stanford: Stanford University Press, p. 209.
74. Ibid.: p. 211.
75. Saxer, Carl J. 2013: "Capabilities and Aspirations: South Korea's Rise as a Middle Power", *Asia Europe Journal*, Vol. 11, No. 4, p. 398.

76. Cha, Victor 2000: "The Security Domain of South Korea's Globalization", in: Samuel S. Kim (ed.), *Korea's Globalization: Challenges and Responses*, Cambridge: Cambridge University Press, p. 236.
77. Ibid.: p. 238.
78. Ibid.: p. 239.
79. Shin, G.W. 2005: p. 624.
80. Myers, Todd Eric 2011: "Asian Values Debate", in: Deen K. Chatterjee (ed.), *Encyclopedia of Global Justice*, Dordrecht/New York: Springer, p. 48.
81. Kim, Dae-jung 1994: "Is Culture Destiny: The Myth of Asia's Anti-Democratic Balues: A Response to Lee Kuan Yew", *Foreign Affairs*, Vol. 73, No. 6, p. 189.
82. Ibid.
83. Snyder, Scott A. and Glosserman, Brad 2015: *The Japan-South Korea Identity Clash: East Asian Security and the United States*, New York: Columbia University Press, p. 64.
84. Ibid.: p. 67.
85. Cho, Il Hyun and Park, Seo-Hyun 2014: "Domestic Legitimacy Politics and Varieties of Regionalism in East Asia", *Review of International Studies*, Vol. 40, No. 3, p. 601.
86. See also John, Jojin V. 2013: "Segyehwa to Global Korea: Globalization and the Shaping of South Korean Foreign Policy", *Area Studies*, Vol. 7, No. 2, p. 17.
87. Kim, Byung-Kook 2006: "The Politics of National Identity: The Rebirth of Ideology and Drifting Foreign Policy in South Korea", in: Jonathan D. Pollack (ed.), *Korea: The East Asian Pivot*, Newport, RI: Naval War College Press, p. 81.
88. Ryu, Yongwook 2013: "Does Political Generation Matter for Foreign Policy? South Korea's Foreign Policy Changes, 2003 to 2007", *The Korean Journal of International Studies*, Vol. 11, No. 1, pp. 117–118.
89. Ibid.
90. Snyder, S.A. and Glosserman, B., 2015: p. 69.
91. Cho, I.H. and Park, S.H. 2014: p. 584.
92. Ibid.: p. 583.
93. Shin, G.W. 2005: p. 625.
94. Campbell, Emma 2011: "Changing South Korea: Issues of Identity and Reunification in Formulating the Australia-Korea Security Policy, Foreign Policy, and Wider Relationship", *Korea Observer*, Vol. 42, No. 1, pp. 117–144.
95. Denney, Steven and Friedhoff, Karl 2013: "South Korea and a New Nationalism in an Era of Strength and Prosperity", *Center for Strategic and International Studies PacNet Newsletter 75*, 7 October. Available at: https://csis-prod.s3.amazonaws.com/s3fs-public/legacy_files/files/publication/Pac1375_0.pdf [Accessed 5 November 2018].
96. Campbell, E. 2011: p. 128.
97. Kim, Youna 2013: "Introduction: Korean Media in a Digital Cosmopolitan World", in: Youna Kim (ed.), *The Korean Wave: Korean Media Go Global*, London/New York: Routledge, p. 1.
98. Ibid.: p. 4.
99. Campbell, Emma 2010: "South Korea's G-Generation: A Nation within a Nation, Detached from Unification", *East Asia Forum*, 13 April. Available at: www.eastasiaforum.org/2010/04/13/south-koreas-g-generation-a-nation-within-a-nation-detached-from-unification/ [Accessed 5 November 2018].
100. Denney, S. and Friedhoff, K. 2013.
101. Ibid.
102. Hermanns, H. 2006: p. 7.
103. Kim, A. 2011: p. 10.
104. Rozman, G. 2012: p. 64.
105. Ducke, Isa 2002: *Status Power: Japanese Foreign Policy Making toward Korea*, New York/London: Routledge, p. 121.

106. John, J.V. 2013: p. 20.
107. See also Snyder, Scott A. 2012: "South Korea's Small Think with Japan", *Asia Unbound Blog, Council on Foreign Relations*, 15 August. Available at: http://blogs.cfr.org/asia/2012/08/15/south-koreas-small-think-with-japan/ [Accessed 5 November 2018].
108. Rozman, G. and Kim, A. 2012: p. 215.
109. Cho, I.H. and Park, S.H. 2014: p. 599.
110. Mo, Jongryn (ed.) 2015: *MIKTA, Middle Powers, and New Dynamics of Global Governance: The G20's Evolving Agenda*, New York: Palgrave Macmillan.
111. Oh, Kongdan 1999: "South Korea's Foreign Policy: A Dolphin among Whales?", *Paper presented to the 14th Annual Conference of the Council on US-Korean Security Studies*, 27–30 October, Arlington, VA; Brazinsky, Gregg 2007: *Nation Building in South Korea: Koreans, Americans, and the Making of a Democracy*, Chapel Hill: University of North Carolina Press; Tudor, Daniel 2014: "Korea Is a Dolphin, Not a Shrimp", *Korea Joongang Daily*, 3 February. Available at: http://mengnews.joins.com/view.aspx?aId=2984326 [Accessed 5 November 2018].
112. Chun, Chaesung 2013: "South Korea's Foreign Policy and East Asia", in: Rüdiger Frank and John Swenson-Wright (eds.), *Korea and East Asia: The Stony Road to Collective Security*, Leiden: Brill, p. 164.
113. Ibid.
114. Ibid.: p. 169.
115. Lee, Sook-jong 2011: "Geullobeoljeongcheseonggwa hangugui oegyo" [Global Identity and Korea's Diplomacy], in: Kang Won-Taek and Lee Nae-young (eds.), *Hangugin, urineun nuguinga? Yeoronjosareul tonghae bon hanguginui jeongcheseong [Understanding Korean Identity: Through the Lens of Opinion Survey]*, Seoul: Dongasiayeonguwon [East Asia Institute], p. 91.
116. Lee, S.J. 2011: p. 92; Lee, Sook-jong 2012: "South Korea as New Middle Power Seeking Complex Diplomacy", *EAI Asia Security Initiative Working Paper, No. 25*, Seoul: East Asia Institute, p. 20.
117. Lee, Nae-Young and Yoon, In-Jin 2016: *Hanguginui jeongcheseong: byeonhwawa yeonsok, 2005–2015 [South Korean Identity: Change and Continuity, 2005–2015]*, Seoul: Dongasiayeonguwon [East Asia Institute], p. 310.
118. Ibid.: p. 312.
119. Ibid.
120. Ibid.: p. 313.
121. Ibid.: p. 314.
122. Ibid.: p. 313.
123. The discrepancy between the numbers in support for ODA in the Asan Institute survey and the EAI survey may simply be explained by the slightly different questions asked. While the EAI asked about the support for an increase in ODA, the Asan Institute asked about general support.
124. Kim, J. James, Choi, Hyeonjung, Lee, John J. and Kang, Chungku 2017: "The Giving Mind: Analysis of South Korean Public and Elite Attitudes on ODA", *Asan Institute for Policy Studies Report*, 24 January, p. 19. Available at: http://en.asaninst.org/contents/the-giving-mind-analysis-of-south-korean-public-and-elite-attitudes-on-oda/ [Accessed 5 November 2018].
125. Ibid.: p. 20.
126. See also Lee, Sook-jong 2015: "South Korea's Middle Power Activism and the Retooling of Its Public Diplomacy", in: Jan Melissen and Yul Sohn (eds.), *Understanding Public Diplomacy in East Asia: Middle Powers in a Troubled Region*, New York: Palgrave Macmillan, p. 117.
127. Lee, S.J. 2015: p. 118.
128. Shin, Gi-Wook and Choi, Joon Nak 2009: "Paradox or Paradigm? Making Sense of Korean Globalization", in: Chang Yun-shik, Hyun-Ho Seok and Donald L.

Baker (eds.), *Korea Confronts Globalization*, London/New York: Routledge, pp. 250–272.

129. Kalinowski, Thomas and Cho, Hyekyung 2012: "Korea's Search for a Global Role between Hard Economic Interests and Soft Power", *The European Journal of Development Research*, Vol. 24, No. 2, p. 252.
130. Ikenberry, G. John and Mo, Jongryn 2013: *The Rise of Korean Leadership, Emerging Powers and International Order*, New York: Palgrave Macmillan, pp. 164–165.
131. Ibid.: p. 8.
132. Hermanns, H. 2006: p. 7.
133. Jordaan, Eduard 2003: "The Concept of a Middle Power in International Relations: Distinguishing between Emerging and Traditional Middle Powers", *Politikon*, Vol. 30, No. 2, pp. 165–181.
134. Kim, Sunhyuk and Lim, Wonhyuk 2007: "How to Deal with South Korea", *The Washington Quarterly*, Vol. 30, No. 2, pp. 71–82; Rozman, Gilbert 2006b: "Regionalism in Northeast Asia: Korea's Return to Center Stage", in: Charles K. Armstrong, Gilbert Rozman, Samuel S. Kim and Stephen Kotkin (eds.), *Korea at the Center, Dynamics of Regionalism in Northeast Asia*, Armonk, NY: M.E. Sharpe, pp. 151–166.
135. See Hundt, David 2011: "Middle Powers and the Building of Regional Order: Australia and South Korea Compared", *Korea Observer*, Vol. 42, No. 1, pp. 69–94; Ikenberry, G.J. and Mo, J.R. 2013; Saxer, C.J. 2013; Kim, Sangbae 2014: "Roles of Middle Power in East Asia: A Korean Perspective", *East Asia Institute MPDI Working Paper*, No. 2, 29 January. Available at: www.eai.or.kr/data/bbs/eng_report/20140203158563.pdf [Accessed 5 November 2018].
136. Nye, Joseph S. 2009: "South Korea's Growing Soft Power", *Project Syndicate*, 10 November. Available at: www.project-syndicate.org/commentary/south-korea-s-growing-soft-power?barrier=accessreg [Accessed 5 November 2018].
137. Chun, C.S. 2013: p. 155.
138. Saxer, C.J. 2013: pp. 398–400; Robertson, Jeffrey 2017: "Middle-Power Definitions: Confusion Reigns Supreme", *Australian Journal of International Affairs*, Vol. 71, No. 4, pp. 355–370.
139. Cooper, Andrew F. (ed.) 1997: *Niche Diplomacy: Middle Powers after the Cold War*, London: Macmillan; Easley, Leif-Eric 2012: "Middle Power National Identity? South Korea and Vietnam in US-China Geopolitics", *Pacific Focus*, Vol. 27, No. 3, pp. 421–442.
140. See also Maass, Matthias M. and Chong, Alan 2010: "Introduction: The Foreign Policy Power of Small States", *Cambridge Review of International Affairs*, Vol. 23, No. 3, pp. 381–382.
141. Cooper, Andrew F. 2015: "Soft Power and the Recalibration of Middle Powers: South Korea as an East Asian Leader and Canada as the Exemplar of the Traditional Model", in: Jan Melissen and Yul Sohn (eds.), *Understanding Public Diplomacy in East Asia: Middle Powers in a Troubled Region*, New York: Palgrave Macmillan, p. 32.
142. Rozman, Gilbert 2006a: "Introduction", in: Charles K. Armstrong, Gilbert Rozman, Samuel S. Kim and Stephen Kotkin (eds.), *Korea at the Center, Dynamics of Regionalism in Northeast Asia*, Armonk, NY: M.E. Sharpe, p. ix.
143. Kim, Sung-Mi 2016: "South Korea's Middle-Power Diplomacy: Changes and Challenges", *Research Paper, Chatham House Asia Programme*, p. 5. Available at: www.chathamhouse.org/sites/files/chathamhouse/publications/research/2016-06-22-south-korea-middle-power-kim.pdf [Accessed 5 November 2018].
144. Ibid.: p. 14.
145. Son, K.Y. 2011.
146. Easley, L.E. 2012.

147. Ibid.
148. See also Koh, B.C. 2001: p. 232.
149. Easley, L.E. 2012: p. 437.
150. Kim, S.M. 2016: p. 3.
151. Ibid.: p. 12.
152. Rozman, G. 2012.
153. Kim, Samuel S. 2006: *The Two Koreas and the Great Powers*, Cambridge: Cambridge University Press, p. 27.
154. See also Ingebritsen, Christine, Neumann, Iver B., Gstöhl, Sieglinde and Beyer, Jessica (eds.) 2006: *Small States in International Relations*, Seattle: University of Washington Press.
155. Woo, Jung-en 1991: *Race to the Swift: State and Finance in Korean Industrialization*, New York: Columbia University Press, p. 9.
156. Ibid.: p. 8.
157. See also Shim, David and Flamm, Patrick 2013: "Rising South Korea: A Minor Player or a Regional Power?", *Pacific Focus*, Vol. 28, No. 3, pp. 384–410.

References

Alford, C. Fred 1999: *Think No Evil: Korean Values in the Age of Globalization*, Ithaca: Cornell University Press.

Brazinsky, Gregg 2007: *Nation Building in South Korea: Koreans, Americans, and the Making of a Democracy*, Chapel Hill: University of North Carolina Press.

Breuker, Remco E. 2009: "Korea's Forgotten War: Appropriating and Subverting the Vietnam War in Korean Popular Imaginings", *Korean Histories*, Vol. 1, No. 1, pp. 36–59.

Campbell, Emma 2010: "South Korea's G-Generation: A Nation within a Nation, Detached from Unification", *East Asia Forum*, 13 April. Available at: www.eastasiaforum.org/2010/04/13/south-koreas-g-generation-a-nation-within-a-nation-detached-from-unification/ [Accessed 5 November 2018].

Campbell, Emma 2011: "Changing South Korea: Issues of Identity and Reunification in Formulating the Australia-Korea Security Policy, Foreign Policy, and Wider Relationship", *Korea Observer*, Vol. 42, No. 1, pp. 117–144.

Cha, Victor 2000: "The Security Domain of South Korea's Globalization", in: Samuel S. Kim (ed.), *Korea's Globalization: Challenges and Responses*, Cambridge: Cambridge University Press, pp. 217–240.

Cho, Il Hyun and Park, Seo-Hyun 2014: "Domestic Legitimacy Politics and Varieties of Regionalism in East Asia", *Review of International Studies*, Vol. 40, No. 3, pp. 583–606.

Chun, Chaesung 2013: "South Korea's Foreign Policy and East Asia", in: Rüdiger Frank and John Swenson-Wright (eds.), *Korea and East Asia: The Stony Road to Collective Security*, Leiden: Brill, pp. 155–180.

Cooper, Andrew F. (ed.) 1997: *Niche Diplomacy: Middle Powers after the Cold War*, London: Palgrave Macmillan.

Cooper, Andrew F. 2015: "Soft Power and the Recalibration of Middle Powers: South Korea as an East Asian Leader and Canada as the Exemplar of the Traditional Model", in: Jan Melissen and Yul Sohn (eds.), *Understanding Public Diplomacy in East Asia: Middle Powers in a Troubled Region*, New York: Palgrave Macmillan, pp. 31–50.

Cumings, Bruce 2005: *Korea's Place in the Sun: A Modern History*, New York: W. W. Norton Company.

De Ceuster, Koen 2005: "Pride and Prejudice in South Korea's Foreign Policy", *Copenhagen Journal of Asian Studies*, Vol. 21, pp. 64–90.
Denney, Steven and Friedhoff, Karl 2013: "South Korea and a New Nationalism in an Era of Strength and Prosperity", *Center for Strategic and International Studies PacNet Newsletter 75*, 7 October. Available at: https://csis-prod.s3.amazonaws.com/s3fs-public/legacy_files/files/publication/Pac1375_0.pdf [Accessed 5 November 2018].
Ducke, Isa 2002: *Status Power: Japanese Foreign Policy Making toward Korea*, New York/London: Routledge.
Easley, Leif-Eric 2012: "Middle Power National Identity? South Korea and Vietnam in US-China Geopolitics", *Pacific Focus*, Vol. 27, No. 3, pp. 421–442.
Eckert, Carter J., Lee, Ki-baik, Lew, Ick, Young, Robinson, Michael and Wagner, Edward W. 1990: *Korea Old and New: A History*, Seoul: Ilchokak Publishers.
Em, Henry H. 2013: *The Great Enterprise: Sovereignty and Historiography in Modern Korea*, Durham, NC: Duke University Press.
Griffis, William Elliot 1882: *Corea: The Hermit Nation*, New York: Scribner.
Ha, Young-Sun 1999: "The Historical Development of Korean Globalization: Kukchehwa and Segyehwa", in: Chung-in Moon and Jongryn Mo (eds.), *Democratization and Globalization in Korea: Assessments and Prospects*, Seoul: Yonsei University Press, pp. 159–178.
Hahm, Chaibong and Kim, Seog-gun 2006: "Remembering Japan and North Korea: Politics of Memory in South Korea", in: Gerrit W. Gong (ed.), *Memory and History in East and Southeast Asia: Issues of Identity in International Relations*, Washington, DC: CSIS, pp. 101–112.
He, Baogang 1999: "The Role of Civil Society Defining the Boundary of a Political Community: The Cases of South Korea and Taiwan", *Asian Studies Review*, Vol. 23, No. 1, pp. 27–48.
Hermanns, Heike 2006: "The Korean Nationalist Discourse Online: An Example of Public Diplomacy in the Internet Era?", *Conference Paper of the 16th Biennial Conference of the Asian Studies Association of Australia at the University of Wollongong*, 26–29 June.
Hundt, David 2011: "Middle Powers and the Building of Regional Order: Australia and South Korea Compared", *Korea Observer*, Vol. 42, No. 1, pp. 69–94.
Hundt, David and Bleiker, Roland 2007: "Reconciling Colonial Memories in Korea and Japan", *Asian Perspective*, Vol. 31, No. 1, pp. 61–91.
Hwang, Balbina Y. 2012: "Korea and PKO: Is Korea Contributing to Global Peace?", in: Scott A. Snyder (ed.), *Global Korea: South Korea's Contributions to International Security*, New York/Washington, DC: Council on Foreign Relations, pp. 13–27.
Ikenberry, G. John and Mo, Jongryn 2013: *The Rise of Korean Leadership, Emerging Powers and International Order*, New York: Palgrave Macmillan.
Ingebritsen, Christine, Neumann, Iver B., Gstöhl, Sieglinde and Beyer, Jessica (eds.) 2006: *Small States in International Relations*, Seattle: University of Washington Press.
John, Jojin V. 2013: "Segyehwa to Global Korea: Globalization and the Shaping of South Korean Foreign Policy", *Area Studies*, Vol. 7, No. 2, pp. 1–28.
Jordaan, Eduard 2003: "The Concept of a Middle Power in International Relations: Distinguishing between Emerging and Traditional Middle Powers", *Politikon*, Vol. 30, No. 2, pp. 165–181.
Kalinowski, Thomas and Cho, Hyekyung 2012: "Korea's Search for a Global Role between Hard Economic Interests and Soft Power", *The European Journal of Development Research*, Vol. 24, No. 2, pp. 242–260.

Kang, David C. 2010: *East Asia before the West: Five Centuries of Trade and Tribute*, New York: Columbia University Press.

Kang, Sung-Hack 2011: *Korea's Foreign Policy Dilemmas: Defining State Security and the Goal of National Unification*, Kent: Global Oriental.

Kim, Andrew 2011: "South Korean National Identity and Its Strategic Preferences", *Joint U.S.-Korea Academic Studies: Korea's Domestic Policies and Their Influence on Asia, Emerging Voices*, Vol. 22, pp. 7–17. Available at: www.keia.org/sites/default/files/publications/emergingvoices_final_andrewkim.pdf [Accessed 5 November 2018].

Kim, Byung-Kook 2006: "The Politics of National Identity: The Rebirth of Ideology and Drifting Foreign Policy in South Korea", in: Jonathan D. Pollack (ed.), *Korea: The East Asian Pivot*, Newport, RI: Naval War College Press, pp. 79–122.

Kim, Dae-jung 1994: "Is Culture Destiny: The Myth of Asia's Anti-Democratic Values: A Response to Lee Kuan Yew", *Foreign Affairs*, Vol. 73, No. 6, p. 189.

Kim, Hun Joon 2011: "Political Legitimacy in South Korea", in: John Kane, Hui-Chieh Loy and Haig Patapan (eds.), *Political Legitimacy in Asia: New Leadership Challenges*, New York: Palgrave Macmillan, pp. 217–237.

Kim, J. James, Choi, Hyeonjung, Lee, John J. and Kang, Chungku 2017: "The Giving Mind: Analysis of South Korean Public and Elite Attitudes on ODA", *Asan Institute for Policy Studies Report*, 24 January. Available at: http://en.asaninst.org/contents/the-giving-mind-analysis-of-south-korean-public-and-elite-attitudes-on-oda/ [Accessed 5 November 2018].

Kim, Key-hiuk 1980: *The Last Phase of the East Asian World Order: Korea, Japan, and the Chinese Empire, 1860–1882*, Berkeley: University of California Press.

Kim, Samuel S. 2006: *The Two Koreas and the Great Powers*, Cambridge: Cambridge University Press.

Kim, Sangbae 2014: "Roles of Middle Power in East Asia: A Korean Perspective", *East Asia Institute MPDI Working Paper*, No. 2, 29 January. Available at: www.eai.or.kr/data/bbs/eng_report/20140203158563.pdf [Accessed 5 November 2018].

Kim, Sung-Mi 2016: "South Korea's Middle-Power Diplomacy: Changes and Challenges", *Research Paper, Chatham House Asia Programme*. Available at: www.chathamhouse.org/sites/files/chathamhouse/publications/research/2016-06-22-south-korea-middle-power-kim.pdf [Accessed 5 November 2018].

Kim, Sunhyuk and Lim, Wonhyuk 2007: "How to Deal with South Korea", *The Washington Quarterly*, Vol. 30, No. 2, pp. 71–82.

Kim, Yong-koo 2001: *The Five Years' Crisis, 1866–1871: Korea in the Maelstrom of Western Imperialism*, Incheon: Circle Books.

Kim, Youna 2013: "Introduction: Korean Media in a Digital Cosmopolitan World", in: Youna Kim (ed.), *The Korean Wave: Korean Media Go Global*, London/New York: Routledge, pp. 1–28.

Kim Haboush, JaHyun 2001: *The Confucian Kingship in Korea*, New York: Columbia University Press.

Koh, Byung Chul 2001: "The Foreign and Unification Policies of the Republic of Korea", in: Kil Soong Hoom and Moon Chung-in (eds.), *Understanding Korean Politics: An Introduction*, Albany: State University of New York Press, pp. 231–268.

Lee, Ki-baek 1984: *A New History of Korea*, Seoul: Ilchokak Publishers.

Lee, Nae-Yong and Yoon, In-Jin 2016: *Hanguginui jeongcheseong: byeonhwawa yeonsok, 2005–2015 [South Korean Identity: Change and Continuity, 2005–2015]*, Seoul: Dongasiayeonguwon [East Asia Institute].

Lee, Sook-jong 2011: "Geullobeoljeongcheseonggwa hangugui oegyo [Global Identity and Korea's Diplomacy]", in: Kang Won-Taek and Lee Nae-young (eds.), *Hangugin, urineun nuguinga? Yeoronjosareul tonghae bon hanguginui jeongcheseong [Understanding Korean Identity: Through the Lens of Opinion Survey]*, Seoul: Dongasiayeonguwon [East Asia Institute], pp. 81–97.

Lee, Sook-jong 2012: "South Korea as New Middle Power Seeking Complex Diplomacy", *EAI Asia Security Initiative Working Paper, No. 25*, Seoul: East Asia Institute.

Lee, Sook-jong 2015: "South Korea's Middle Power Activism and the Retooling of Its Public Diplomacy", in: Jan Melissen and Yul Sohn (eds.), *Understanding Public Diplomacy in East Asia: Middle Powers in a Troubled Region*, New York: Palgrave Macmillan, pp. 107–129.

Maass, Matthias M. and Chong, Alan 2010: "Introduction: The Foreign Policy Power of Small States", *Cambridge Review of International Affairs*, Vol. 23, No. 3, pp. 381–382.

Mishra, Sandip Kumar 2011: "Korean National Identity Discourse: Evolution, Colonialism, and Division", *The Journal of Northeast Asian History*, Vol. 8, No. 2, pp. 5–33.

Mo, Jongryn (ed.) 2015: *MIKTA, Middle Powers, and New Dynamics of Global Governance: The G20's Evolving Agenda*, New York: Palgrave Macmillan.

Moon, Chung-in and Li, Chun-fu 2010: "Reactive Nationalism and South Korea's Foreign Policy on China and Japan: A Comparative Analysis", *Pacific Focus*, Vol. 25, No. 3, pp. 331–355.

Moon, Chung-in and Suh, Seung-won 2005: "Security, Economy, and Identity Politics between Japan and South Korea under the Kim Dae-jung Government", *Korea Observer*, Vol. 36, No. 4, pp. 561–602.

Myers, Brian R. 2010: *The Cleanest Race: How North Koreans See Themselves and Why It Matters*, Plymouth: Melville House Publishing.

Myers, Todd Eric 2011: "Asian Values Debate", in: Deen K. Chatterjee (ed.), *Encyclopedia of Global Justice*, Dordrecht/New York: Springer, pp. 48–50.

Noland, Marcus, Robinson, Sherman and Wang, Tao 2001: "Famine in North Korea: Causes and Cures", *Economic Development and Cultural Change*, Vol. 49, No. 4, pp. 741–767.

Nye, Joseph S. 2009: "South Korea's Growing Soft Power", *Project Syndicate*, 10 November. Available at: www.project-syndicate.org/commentary/south-korea-s-growing-soft-power?barrier=accessreg [Accessed 5 November 2018].

Oh, Kongdan 1999: "South Korea's Foreign Policy: A Dolphin among Whales?", *Paper presented to the 14th Annual Conference of the Council on US-Korean Security Studies*, 27–30 October, Arlington, VA.

Phillips, Andrew 2018: "Contesting the Confucian Peace: Civilization, Barbarism and International Hierarchy in East Asia", *European Journal of International Relations*, Vol. 24, No. 4, pp. 740–764.

Robertson, Jeffrey 2017: "Middle-Power Definitions: Confusion Reigns Supreme", *Australian Journal of International Affairs*, Vol. 71, No. 4, pp. 355–370.

Robinson, Michael E. 1984: "National Identity and the Thought of Sin Ch'aeho: Sadaejuŭi and Chuch'e in History and Politics", *Journal of Korean Studies*, Vol. 5, pp. 121–142.

Rozman, Gilbert 2006a: "Introduction", in: Charles K. Armstrong, Gilbert Rozman, Samuel S. Kim and Stephen Kotkin (eds.), *Korea at the Center, Dynamics of Regionalism in Northeast Asia*, Armonk, NY: M.E. Sharpe, pp. ix–xiii.

Rozman, Gilbert 2006b: "Regionalism in Northeast Asia: Korea's Return to Center Stage", in: Charles K. Armstrong, Gilbert Rozman, Samuel S. Kim and Stephen Kotkin (eds.), *Korea at the Center, Dynamics of Regionalism in Northeast Asia*, Armonk, NY: M.E. Sharpe, pp. 151–166.

Rozman, Gilbert 2012: "South Korean National Identity: A Six-Dimensional Analysis", in: Gilbert Rozman (ed.), *East Asian National Identities: Common Roots and Chinese Exceptionalism*, Stanford: Stanford University Press, pp. 47–72.

Rozman, Gilbert and Kim, Andrew 2012: "Korean National Identity: Evolutionary Stages and Diplomatic Challenges", in: Gilbert Rozman (ed.), *East Asian National Identities: Common Roots and Chinese Exceptionalism*, Stanford: Stanford University Press, pp. 197–218.

Ryu, Yongwook 2013: "Does Political Generation Matter for Foreign Policy? South Korea's Foreign Policy Changes, 2003 to 2007", *The Korean Journal of International Studies*, Vol. 11, No. 1, pp. 113–141.

Saxer, Carl J. 2013: "Capabilities and Aspirations: South Korea's Rise as a Middle Power", *Asia Europe Journal*, Vol. 11, No. 4, pp. 397–413.

Schmid, Andre 2002: *Korea between Empires, 1895–1919*, New York: Columbia University Press.

Shim, David and Flamm, Patrick 2013: "Rising South Korea: A Minor Player or a Regional Power?", *Pacific Focus*, Vol. 28, No. 3, pp. 384–410.

Shin, Gi-Wook 2005: "Asianism and Korea's Politics of Identity", *Inter-Asia Cultural Studies*, Vol. 6, No. 4, pp. 610–624.

Shin, Gi-Wook 2006: *Ethnic Nationalism in Korea: Genealogy, Politics, and Legacy*, Stanford: Stanford University Press.

Shin, Gi-Wook 2008: "Nation, History, and Politics", in: Hyung Il Pai and Timothy R. Tangherlini (eds.), *Nationalism and the Construction of Korean Identity*, Berkeley: Institute of East Asian Studies, University of California, pp. 148–165.

Shin, Gi-Wook and Choi, Joon Nak 2009: "Paradox or Paradigm? Making Sense of Korean Globalization", in: Chang Yun-shik, Hyun-Ho Seok and Donald L. Baker (eds.), *Korea Confronts Globalization*, London/New York: Routledge, pp. 250–272.

Snyder, Scott A. 2012: "South Korea's Small Think with Japan", *Asia Unbound Blog, Council on Foreign Relations*, 15 August. Available at: http://blogs.cfr.org/asia/2012/08/15/south-koreas-small-think-with-japan/ [Accessed 5 November 2018].

Snyder, Scott A. and Glosserman, Brad 2015: *The Japan-South Korea Identity Clash: East Asian Security and the United States*, New York: Columbia University Press.

Son, Key-young 2011: "From a Garrison State to a Humanitarian Power? Security Identities, Constitutive Norms and South Korea's Overseas Troop Dispatches", *The Korean Journal of Defense Analysis*, Vol. 23, No. 4, pp. 557–573.

Tudor, Daniel 2014: "Korea Is a Dolphin, Not a Shrimp", *Korea Joongang Daily*, 3 February. Available at: http://mengnews.joins.com/view.aspx?aId=2984326 [Accessed 5 November 2018].

Woo, Jung-en 1991: *Race to the Swift: State and Finance in Korean Industrialization*, New York: Columbia University Press.

3 Identity and Roles in IR

Identity is one of the main concepts in IR, and its prominence has much to do with the emergence of the interpretative and postrationalist approach of constructivism since the 1990s. However, due its inherent complexity, there have been a great diversity of different ontological and epistemological assumptions employed in the study of identity, leading to infamous "definitional anarchy."[1] Recently there have been attempts in IR to coordinate and integrate different identity conceptualizations in order to realize the potential of identity-based research while acknowledging the plurality of identity perspectives. After a brief discussion of the developments and shortcomings in constructivist identity research, some parameters are gathered for what a theoretically sound and empirically useful identity approach has to encompass and, from this, the integrative potential of role theory as a further way forward is introduced.

Identity Theories

Identity as a concept in IR is an analytically elusive and contested issue. Because of its multiple applications and its conceptual vagueness, some social scientists[2] have even argued for abandoning the concept completely. However, if vagueness is a problem for complex phenomena, perhaps IR scholars also have to abandon concepts like power or the state.[3] Identity has been explored on various levels of analysis, from the individual to social groups, states, and international organizations and from different sociological, psychological, and philosophical traditions. This can lead to a great deal of confusion, as identity may refer to the individual beliefs and actions of a state leader, the politics of belonging and citizenship in a political community, or the role and interests of an international military alliance such as the North Atlantic Treaty Organization (NATO). The status of identity, "*the* shooting star of international relations theory,"[4] is closely linked to the emergence of what became known as "constructivism" in the late 1980s and early 1990s. Before, identity did not feature greatly in the traditional rationalist theories of realism and liberalism, as both approaches treated the state as a more or less autonomous unit in an anarchical systemic environment, which was seen as either pitting states against each for survival or incentivizing them to overcome this conflictual setting by cooperation, facilitated by mutual trust.

For (neo)realists, all states have the same basic interest in survival and so they fight for the better relative gains in the zero-sum game that is international politics. According to this understanding, state identities are fixed, systemically determined, and states are self-help units.[5] A further specification is maybe the categorization of state actors in super, great, middle, or small powers as measured by their material power capabilities that equips those states with different options for action. As in the case of the ROK during the Cold War, a small power, hence, is likely to balance, bandwagon, hedge, or counterbalance together with other states against a greater power. And only weak or failed states have not enjoyed sovereignty, which is the historical lesson for realists.[6] Because (neo)realists solely focus on systemic determinants for state behavior, they not only neglect the domestic debating and decision-making but also call for strategic and rational thinking according to those systemic determinants.

Proponents of the "liberal school" of IR thought share the basic assumption about the ontology of the international sphere and about state identities, but they draw a different conclusion from that: cooperation—action, not identity—actually helps to overcome the zero-sum situation as long as state actors concentrate on absolute gains.[7] Furthermore, (neo)liberals have a tendency to ascribe different characteristics to different regime types: democracies are said to behave differently from nondemocracies as, for example, illustrated in the liberal peace theory, which posits that democracies are less likely to fight wars with each other. This perspective typically focuses on the benefits of interdependence through trade, international institutions, and regional integration. The key property of sovereignty here remains crucial, but can actually be compromised, if for rational, absolute-gains considerations. According to Harris, "'the nation' and its identity do not appear to feature highly on the agenda that deals with 'real' national interest analyzed by classical rationalist theories of international relations."[8] Constructivists, however, disagree with the rationalist ontology of the axiomatic status of a stable anarchical international system, emphasizing the dynamic social construction and co-constitution of structure and agents through state identities or, in the words of Wendt, "[a]narchy is what states make of it."[9]

When the nascent constructivist approach was challenged by the rationalist establishment of the discipline to adopt a positivist epistemology in order to develop into a "proper scientific" approach, a split between constructivists was the result.[10] Those answering the call to come up with testable hypotheses in order to "be seriously considered [as] a contender in IR theory"[11] started focusing on norms as their prime research object and worked, for example, on national security[12] or the social construction of national interest through interaction.[13] At the same time, more radical constructivists continued to resist these "disciplining expectations"[14] and pursued critical projects on, for example, discourses of national security[15] or the constitution of collective identities.[16] In consequence, identity research ended up being rather fragmented. Whereas the norm literature focused mostly on the diffusion of norms through socialization of actors, poststructural identity perspectives focused most of their attention on processes of "Othering."

Furthermore, Zehfuss criticizes the most prominent constructivist take on identity, Wendt's *Social Theory of International Relations*,[17] for failing to account for the domestic formation of identity and simply treating the state as a unitary actor. By doing so, Wendt takes a systemic perspective and simply looks at changes in behavior while attributing them to changes in identity. Such an approach reduces "identity" to a mere behavioral profile and "to something negotiable between states."[18] Cedermann and Daase thus called for constructivists to "endogeniz[e] corporate identities."[19] By corporate identity they mean the pre-interaction identity of a state actor, "constituted by [. . .] territory, legal frameworks and other institutions,"[20] in contrast to social identity, such as a small or great power. Besides, Aggestam's take on foreign policy and identity is representative for the early "culture"-oriented constructivism in the 1990s. She treated "culture" as the "broader context in which individual and collective identities are linked, producing shared meanings that influence the framing of political action"[21] and, as a consequence, conflating identity with the related concept of "role."

McCourt criticized this widespread conflation of the two concepts, which led in effect to the results of "multiple 'social identities'"[22] of state actors, for example, in the works of Legro[23] and Hopf.[24] On the distinction between identity and role, McCourt argues that most constructivist identity approaches simply assume a direct link between identity and action, according to which "states act because their identities predispose them to certain interests or preferences."[25] For him, because agency is not simply following out of a given identity, such identity-based accounts "leave open the important question of why observed action was socially meaningful in any particular instance, which requires recourse to social roles."[26] If not, actors would simply be slaves of their defining properties, without any real agency to themselves. According to McCourt then, "roles are [. . .] the necessary social vehicle for action in its meaning-creating, identity-affirming sense."[27] This approach is also followed in the conceptualization of identity of this book. More on role-theoretical approaches to international identity and agency is discussed later in the chapter.

Similar to the constructivist split in IR, which is mostly concerned with questions about the international system, order, or society, there are more positivist and more interpretative identity perspectives[28] in the subfield FPA, which focuses on understanding and explaining the behavior of specific actors within the system by unpacking the unitary actor's foreign policy making. Whereas FPA positivists are more concerned with questions of fixed identities and their effects on foreign policies—"causes for action"—the interpretative strand emphasizes the crucial role of language as a medium for the social constitution of identities: how it was possible that a certain identity understanding was established discursively by certain agents—"reasons for action."[29] Common to all identity-based FPA is the focus on effects of national self-understandings on interest formation and consequently on foreign policy action, mostly for the purpose of either understanding foreign policy behavior or the formation of the international system.

Despite the multitude of different usages and definitions of especially the concepts of "identity," "culture," and "role," there actually seems to be a pattern underlying most of the prevalent approaches, mirroring the distinction between "I" and "Me" in the formation of the "Self," which was introduced by American pragmatist philosopher and sociologist George Herbert Mead.[30] For example, the first, more fundamental and pre-interaction part of the Self, has been termed "corporate identity" by Cedermann and Daase or "role identity" by Wendt and, more broadly, "culture" by Aggestam, whereas the latter, the social and contextual manifestation of the Self, has been defined as either "social identity" (Cedermann/Daase), "role" (Wendt), or just "identity" (Aggestam).

Moreover, the agreement by most constructivists on the relational and contextual nature of identity led to the adoption of a notion from sociology and social psychology about "multiple identities." Following McCourt,[31] it is worthwhile to problematize the application of this idea to polities in particular. If one accepts that the relational and contextual nature of "identity" may lead to different social identities, we can indeed speak of multiple (social) identities. But how far does this multiplicity go? Does it affect the more stable, elusive, and fundamental Meadian "I" or "corporate identity"? If so, what ties the multiplicity of identities ultimately together, what is the referent object? If a polity engages in various social contexts and develops a multitude of identities that are anything more than social, does this not automatically lead to the fragmentation and dissolution of the initial political community into a multitude of political communities? Must there not be a more fundamental corporate identity part that experiences and ensures the sameness throughout time and space and, thus, various social contexts, even though it is certainly not immune to changes influenced by the social environment? For this reason, the approach here is following McCourt and Harnisch,[32] in that the ROK has only one corporate identity as a state, which may be manifested in different social or role identities in different international contexts. More on this perspective will follow in Chapter 4, where the methodological approach is presented.

In his entry on identity in the *Handbook for International Studies*, Berenskoetter offers the most comprehensive review of the concept from an IR perspective to date.[33] Acknowledging the complexity of identity, which encompasses the individual and the collective and the state and the nation, as well as historical, cultural, and material aspects, he rebuffs simple or parsimonious conceptualizations of identity and makes the case for a set of parameters that constitute the core of any theoretically rigorous and empirically applicable understanding of identity in IR. According to Berenskoetter, there is a commonsense agreement that "having an identity means having a sense of self,"[34] without which the actor would be unable to perceive and enact any meaningful agency. This precarious state has been termed "ontological insecurity,"[35] and the need for a stable sense of self on the individual level has been attributed to a human desire for cognitive stability[36] and for belonging and recognition.[37] The challenge for constructivism is, however, how to understand the formation of this inherently dynamic and situational idea about oneself.

Here, drawing from works by Hall and Bloom[38] on the desire for a stable sense of self, Berenskoetter introduces the satisfaction of a "will-to-manifest-identity" as the main driving force behind social agency: "what one 'is' (or wants to be) is sustained by what one 'does.'"[39] This way he highlights the performative aspect of identity and the significance of active processes of identification. Identity here is a dynamic, situational, and nonessential phenomenon that must be constantly reaffirmed through performance and can thus only be stabilized temporarily. Herein lies also the difference between the broader and fuzzier term *culture*, which he dubs a "depository of various structural parameters in a society,"[40] and *identity*, "the action unit of culture,"[41] which is established through the selective invocation of certain cultural parameters, generally understood as identity politics. Last, Berenskoetter notes the central role of drawing boundaries with Others, against which or with which the Self is formed. This last aspect, known as "Othering," is at the core of a poststructural understanding of state identity, in which the constitution of difference is seen as the main practice of foreign policy.[42] But because of this "neglect [of] non-hostile implications of self and other distinctions,"[43] the result is an overemphasis of just one dimension of identity and thus on conflict and threats to the Self. Also, many studies invoking national identity tend to focus solely on the politics of drawing boundaries, for example, with regard to multiculturalism, cosmopolitanism, or the emergence and nurturing of a European supranational identity.

Stark Urrestarazu's account of identity as a concept complements Berenskoetter's perspective fittingly. While she also acknowledges the vast variety of different identity perspectives, she criticizes that too often scholars opt for tailored and self-serving *ad hoc* definitions of identity or that they exhibit a tendency to just address partial aspects of identity. In order to yield the "highest possible theoretical exploit"[44] she, too, calls for a multidimensional understanding of identity as an "analytical prism," similar to Berenskoetter's "eye opener" analogy, to shed light on the complexities of the practice and constitution of the actors and structures of world politics. To do so, she proposes a multidimensional model of identity focusing on narratives, performances, and emotions. It should be stated, however, that Berenskoetter and Stark Urrestarazu are speaking of identities of organized collectives, particularly of states and their foreign policies. In contrast to many other identity approaches, most notably Wendt, the state is hence not a unitary actor but a composite actor, a social structure itself.

Starting from the premise that there is no foreign policy practice that is not driven by identity, Stark Urrestarazu highlights the constitutive role of the intersubjective "establishment of continuity via a narrated 'story' of the actor's own historical development as a collective."[45] Berenskoetter, too, elaborates on the narrative side of identity while advocating for the study of biographical narratives in order to take the internal processes of collective identity formation within a given polity seriously,[46] rather than just focusing on Othering. He sees the function of biographical narrative in providing

> the Self with knowledge about its place in 'the world,' specifically to meaningfully situate the Self and delineate its existence in time and space, to provide us with a necessary sense of orientation about *where* we come from and *where* we are, or could be going.[47]

A great number of IR scholars[48] have been applying narrative approaches to identity matters as well as role theory,[49] and the narrative aspects of identity formation are widely accepted, however, with one major caveat: one has to be careful not to essentialize identity again,[50] by condensing fixed and timeless unitary "national characters" or "national cultures." In order to escape this trap, a second identity dimension deserves attention: performance.

In its performative dimension, identity is constituted situationally and relationally, as "actors might advert to several different possible narratives in a given situation."[51] Crucial here is the actual selective identification with certain narratives over others and the consequential actions. This performative dimension is best illustrated through recourse to Mead's work on the Social Self,[52] which was later subsumed under the label "symbolic interactionism" by Mead's student Herbert Blumer and introduced to the social sciences by Erikson.[53] It is in concrete moments of interaction, when the Self is facing an Other, when

> self-fulfilment takes place through an evolutionary process in which the acting Self, the "I" driven by intuition and instinct, gradually comes to internalize the 'attitude of the whole community,' devising a sense of "Me" by adapting to a social group.[54]

This view is also supported by the influential Social Identity Theory (SIT),[55] which presents a theory of action centered on in-group/out-group dynamics. SIT locates the constitution of identity in interaction and ongoing comparison with Others, "resulting in multiple in-group/out-group categorizations based on emotion-laden situational social similarities and differences."[56] Accordingly, deviant in-group identity conceptions constitute a threat but are also the only source for identity change.[57] The unfortunate consequence of such an identity perspective is the rather narrow focusing on binary identity and conflict, neglecting the potentially positive and confidence-building function of identifying with an Other. From a perspective of International Political Sociology, Adler-Nissen argues that SIT approaches tend to neglect the social constitution of the Self in favor of a psychological process vis-à-vis others assuming a universal desire for self-esteem.[58] The more recent literature on status[59] or prestige[60] in IR similarly builds on this psychological understanding of the Self. In consequence, as Adler-Nissen criticizes, SIT approaches fail to note that international positions of state actors can also be "a result of their dealings with their own past and domestic conflicts."[61]

Furthermore, the focus on situational and performative dimensions ties in well with more recent adaptations of relational sociology[62] in the study of international politics. These studies[63] adopt a relational-processualist perspective mostly in a critique of the great body of work on norms in international relations, where "states have the choice either to reject or embrace a given norm but not to (re)define its meaning."[64] Relational-processual approaches stress the role of norms for orientation but insist that actors have creative agency and are not simply norm dupes blindly following a script. Key in this understanding is the theoretical premise that actors only gain consciousness over their actorhood in relation to others due to their sociability. Furthermore, it is through "drawing on norms [that] actors are able to make sense of indeterminate situations with which they are constantly confronted."[65] Thus, proponents of this approach claim to account more fruitfully for norm change through taking agency seriously. Their research strategy thus is to "look at the interpretations, reasons and justifications that agents offer for their actions and to investigate how they change."[66]

Moreover, the situationality of identity results in different identity manifestations through practice, depending on the social context. This also means that foreign policy practice cannot simply be inferred (or predicted) from identity representation retrieved independently from the practice in question, because every action depends on an actual mobilization of identity elements by actors in any given situation.[67] Accordingly, the same actor can mobilize the same identity understanding in different social contexts to a different outcome. Lebow, however, boldly discards the use of the term *identity* for collectives, because for him it implies the existence of a "true" Self, whereas it is "really a composite of numerous self-identifications."[68] His argument rests on the premise that states have no reflexive self, the Meadian "I," as "there is no conscious center able to develop its own identification and resist, assimilate or reformulate those imposed from the outside."[69] For him, roles, affiliations, bodies, and narratives are the main sources for self-identifications, with different identity claims simply imposed on the community by its members. Importantly, Lebow concedes that, when taking an international perspective, states "have identifications that are in part self-chosen in the sense that they are coming from within. Other identifications are external and originate with other actors or are negotiated by states."[70] From this, one can equate his suggested identifications to more conventional terms such as identity conceptions, identity representations, or self-identifications. One wonders whether the solution for the terminological confusion may be in reserving the term identity solely for the elusive sense of Self of the state actor or the corporate identity, as McCourt and Harnisch have argued,[71] thus presuming an ontology of only one identity with differently manifested self-identifications per actor. Also, one would then focus on the easier-to-trace different social dimensions of identity in different social contexts, which are manifested through self-identification practices but nevertheless contested from the inside as well as from the outside (Other-identification). Such a reading is still in line with the gist of Lebow's intervention but does not disregard the use of the term *identity* for collectives or composite actors completely. As will be shown in Chapter 4, this is how identity has been conceptualized in this study.

Third, more and more scholarly attention[72] is dedicated to a rather neglected aspect of identity: emotion. Following social psychological insights, any identity manifestation, (re-)formed through narration and performance, is also emotional.[73] Stark Urrestarazu points to the well-established link between group membership and personal self-esteem, thus highlighting the importance of political status enhancement or humiliation.[74] For some scholars working with psychological insights, experiencing ontological insecurity is *feeling* ontological insecurity.[75] Mercer, for example, differentiates between individual and group emotions, suggesting that identity always depends on social emotions, because "[i]dentification requires a feeling of attachment; it is intrinsically social."[76] But group members also share, validate, and discipline each other's feelings, thus structuring social relations,[77] as otherwise "[a] group without emotion is a mere collection of autonomous individuals."[78] From Mercer's SIT perspective, at least, it appears plausible that the importance of recognition of a Self by Other(s) also stems from its emotional value of positive external affirmation, as suggested by the growing body of work on status in international relations.[79] In contrast, Reus-Smits questions whether a state can indeed be treated as a social group, such as a family, and argues from an institutionalist perspective that the state cannot "be reduced to society: the two must be treated as analytically distinct."[80] While this debate in IR about the role of emotions in world politics has been described as "one of the field's 'great frontiers,'"[81] it remains largely theoretical, with few particularly discursive empirical applications beyond the "'emotions matter' approach of the first wave of emotions scholarship in IR."[82]

Last, Berenskoetter points to the specific situations in which actors encounter each other "with a stock of private knowledge providing a stable sense of Self in time and space with the objective to keep their narrative going."[83] It is this "inter"space between actors where identity is manifested in practice through narration, performance, and emotions that should be the starting point for empirical analyses.[84] Thus, for international politics, the practice of foreign policy is the site where the formation and contestation of international identity take place. It is here where countries' relationship to the outside world, their places in the world, are (re)calibrated.

After this brief review of the identity as a concept in IR theory and FPA, it becomes clear that any identity-focused analysis has to encompass certain parameters and to face specific challenges: in the absence of a consensus on how to conceptualize and identify identity even within constructivism, any identity approach has to make clear the specific operationalization, the level of analysis, whose identity is investigated, where actorhood is seen as located, who is acting on behalf of the identity bearer, and how identity is formed and changed. If one is particularly interested in the identity representations underpinning foreign policy practice of a given state, it seems important to only claim inquiry into the international identity of that state and not the identity or broader culture of that community as a whole, even though these are certainly connected. Furthermore, in order to avoid reductionism, one has to account for more than just one dimension of identity; it is not enough to just rely on either the internalization of norms, narrative representations, or processes of Othering.

From the preceding review, one can conclude that identity, individual and collective, in IR literature is about stabilizing a sense of self that is important for meaningful agency and a sense of belonging. Because identity is contingent and situational, it can only ever be fixed or stabilized temporarily, which highlights the performative dimension of identity as a practice of identification. Thus, any identity research should focus on specific practices of the actor of concern and the related identity politics, because it is only here where an otherwise contingent and broader sense of self becomes actualized in a specific situation to a specific time. For the stabilization of a certain identity understanding through time and space, the narrative dimension stands out, as situational identity representations are bound together to a biographical narrative. Moreover, identity is contested, from within and without. Identification with or against others takes place in order to make sense of "being" in a given situation. Both a logic of appropriateness as well as a logic of consequence can be the result, depending on the creative agency of actors and recognition by others. While there is an interesting IR debate ongoing about social emotions and state identity, this remains still largely theoretical and with little empirical applicability for now. Finally, it should be said that, because of the overly systemic focus of especially the norm-oriented socializing literature, a perspective focusing on internal or domestic identity politics, meaning the formation and contestation of different identity claims in different practice situations, seems warranted.

Role Theory and Its Integrative Potential

With the aforementioned identity parameters in mind, it is worthwhile to introduce the very similar but distinct concept of role and the corresponding theoretical approach. The rich role-theoretical conceptual language can have an integrative function regarding the already discussed identity perspectives, clarifying terminology, and opening an empirically promising way of grasping international state identities beyond their manifestations in social or role identities. To put it very simply, roles are the results of the encounter of an actor's identity with the expectations of a given social environment. Any role must be based in a corresponding identity element, and any identity reaffirmed by respective role-playing. Hence, identity affirmation is only possible due to role-playing, as only social roles give identity meaning.[85]

The sociological concept of role was first introduced into FPA by Holsti in 1970, but early approaches focused mainly on elite perceptions and on the restraining effects of role-based action.[86] For Holsti, role theory offered a framework for describing foreign policy behavior, or "role performance," and role conceptions and their sources.[87] He defines *role* or *role performance* as "behaviour (decisions and actions) [which] can be kept analytically distinct from *role prescriptions*, which are the norms and expectations cultures, societies, institutions, or groups attach to particular *positions*."[88] Still, he contends that because of the relative social thinness of international politics, expectations of Others (*alter*) are less relevant compared to internal role-making processes (*ego*). This is the reason he suggests we speak of the more dynamic status in international politics instead of a more rigid position as in more structured social settings.

According to Harnisch,[89] this *ego*-dominance was typical for the first phase of role-theoretical literature in FPA until the 1990s, when scholars mainly focused on tracing national role conceptions held by state leaders. Walker's comprehensive and systemic stock take of role-theoretical advances[90] also explore the possibilities of role-theoretical insights for the study of international politics during this phase. In the second, ongoing phase scholars started turning more to interaction and the role expectations held by others,[91] investigating role conflict,[92] the challenge for an actor of having to balance different expectations for different held roles, and role socialization,[93] as well as the domestic sources of roles and role contestation.[94] In mainstream constructivism, however, the use of role theory remains largely "implicit"[95] and, for the most part, anecdotal. Wendt, for example, introduces different cultures of international politics according to only three ideal type state identities, rival, friend, or enemy,[96] with this understanding of international roles later being criticized as very static and reductive.[97] Finally, even the most eminent neorealist scholar Kenneth Waltz used the language of role theory with regard to bipolarity, as Thies argues.[98]

It is important to note that there is no single role theory but, rather, different sociological strands with sometimes conflicting use of concepts. Nabers identifies five perspectives, structuralist, functional, symbolic interactionist, organizational, and cognitive role theory, each with their own strengths and shortcomings.[99] While the first two standpoints are seen as overemphasizing the constraining functions of norms, the symbolic interactionist approach, for example, has been criticized for neglecting contextual constraints on actors' interactions. Adler-Nissen, in contrast, argues that "[i]nstead of interpreting interactions as a one-way disciplining process, symbolic interactionism explores the negotiation of social order as a complex interactive process."[100]

Therefore, role theory is also capable of bridging these levels of analysis, the individual and the domestic sphere, as well as the international system. Holsti describes the international system as a distribution of different roles, already foreclosing the constructivist premise about the dynamic nature of the international system. Thus, in reference to Wendt's famous catchphrase, Harnisch, Frank, and Maull coin this insight as follows: "international social order is what states make of it, and thus what roles they play."[101] In consequence, much role-theoretical work has been dedicated to questions about the stability of orders due to different role-playing by the respective actors. There is a great variety of different understandings of roles in the literature such as rival, enemy, friend, superpower, great power, residual great power, leader of the West, middle power, balancer, faithful ally, defender of the faith, status quo–oriented power, robust defender of human rights, nonaligned power, sovereign state, or colonial power. But as roles and thus order emerge from the interaction of actors specific to time and space, there is no agreed-on catalogue of roles. Furthermore, the different role types have different scopes and criteria. Superpower, for example, might also be considered a "role set" or a master role that is supported by auxiliary roles, such as leader of the West and robust defender of human rights.[102]

Most recently, a number of IR scholars[103] have reinvigorated the symbolic interactionist strand in role theory in FPA, making use of the Meadian understanding of individuation as a constant social process between an "I" and a "Me." The latter is the self-image when actors look at themselves through the eyes of the other, through "taking the role of an Other." In addition to significant concrete Others, Mead[104] makes the case for the importance of the generalized other for the "Me," which is anticipating and reflecting upon the internalized, whether assumed or experienced, attitudes in the respective social context, without taking concrete expectations into account.[105] It is through this interpretative process of self-identification, of "taking the role of the other," that the Self is constituted. The sense of self, or corporate self-identity, is thus located in the ego part[106] of which role conceptions are an integral part in the "Me," but not the role itself, which is the concrete, situational manifestation of ego-alter interaction. According to Mead, only this "Me" is empirically observable in a specific action-demanding situation: "We are aware of ourselves, and of what the situation is, but exactly how we will act never gets into experience until after the action takes place."[107] Importantly, the experiencing "I" part is only known through reflection, becoming objectified or manifested as a "historical figure"[108] in the "Me."

The crucial outcome from this perspective is that the self is not a passive recipient of social rules or a zombie-like executor of his basic character, for that matter, but an active interpreter of social attitudes when taking up the role of the generalized other.[109] Here, roles are not understood as detailed scripts that are directing decision-makers' behavior but as the emergent relational outcomes of the ongoing process of interaction, with the "Me" providing a stable or habitual framework for agency and the evasive "I" as a feeling and creative mainspring. Importantly, the "me" can also be a "past me," a historical self, against which or with which an actor self-identifies in the present day.[110] Harnisch, for example, shows this with an analysis of how China's international roles are manifestations of changing historical self-identifications related to colonial victimhood in the context of the "century of humiliation" narrative as well as to modern developmental achievements as an emerging power.[111]

International roles relate to identities in different ways: roles are specific manifestations of ego-alter expectations, whereas identities also include ideas and emotions about the self that are maintained throughout time and space.[112]

In consequence, the past, experience, and memory are all part of an actor's identity as "former stages in the self-identification process."[113] The longer the self-identifications go back and the higher their institutionalization, the stronger the incentives to the related role performance and expectations.[114] Harnisch refers to these constitutive identities as individual or corporate, and he locates the discussion of ontological security here.[115] But he also mentions further identity types, such as social identities stemming from group membership or type identities that refer to specific characteristics, such as being an autocracy or a socialist country. Last, he also defines role identities "as conceptions of individuals or states about themselves as role players."[116] According to Harnisch, only these role identities account for a particular social task or purpose, whereas social or type identities fulfill constitutive functions. Moreover, his understanding of the state as a social structure opens the opportunity for looking into the ego part of an international role, which is open to domestic contestation.[117] Here, role theorists distinguish between vertical and horizontal perspectives on domestic role contestation: whereas the vertical contestation is, for example, between elites and masses, the horizontal is between different political agents within the same polity.

It should be noted that in the role-theoretical literature, there are different understandings of the relationship between role and identity. For Nabers, identity simply equals role identity, which could also be seen as a dominant role conception that is linked to an actually enacted role and therefore not a rivaling, deviant, or counter-role conception.[118] Most of the related literature is primarily interested in exploring the role-playing of a state actor and the negotiation of different expectations while neglecting the related domestic political processes.[119] Thus, Cantir and Kaarbo attempt to unpack the ego with a focus on domestic role contestation. For them, "both identities and roles are ideational factors that have a bearing on a state's behaviour, [with] the latter carry[ing] more specific prescriptions for action."[120]

Now, the formation of role conceptions, which can be abstractly understood as a conversation within the ego between "I" and "Me" in a given situation prior to concrete alter expectations, is where identity elements are utilized in order to identify with the internalized ideas about the generalized other. In the case of state actors, it is fair to assume that this contested decision-making process is about the legitimation of actions in reference to certain identity elements, which are consequently reaffirmed through role performance.[121] It is also possible though that an actor grows increasingly away from a role enactment with decreasing identification in "a sort of detachment [. . .] which goes with dissatisfaction and estrangement."[122] This is called role distance and is a key indicator for role and identity change.[123] It is this link between evasive identity elements that is manifested through the reflective and contested formation of role conceptions and, in turn, lends legitimacy to specific role performances that can serve as the entry point for inquiries into identity.

In both constructivist identity research, as well as in the study of South Korea as a foreign policy actor, role-theoretical rhetoric has been used occasionally but mostly anecdotally and seldom to its full potential.[124] Moreover, role-playing has been often understood in a simplistic way. An illustrative example of the latter is the socialization-focused argument by Checkel that rejects role-playing as superficial internalization and mere simulation of behavior.[125] But because of his focus on norm internalization imposed from the outside, Checkel overlooks internal processes of identity formation.

With regard to the use of role-theoretical language in studies on South Korea, Easley refers to middle power as a role,[126] although against the background of the wider role-theoretical literature, middle power is better seen as a role set. Kim, in contrast, compiles four "identities [which] exemplify the distinctive characteristics of middle power":[127] early mover, bridge, coalition coordinator, and norm diffusor, all of which are obviously better defined as roles or role identities forming together a middle-power role set. Only Hermanns has so far applied role-theory concepts directly for an analysis of Lee Myung-bak's Global Korea policy.[128] She derives three different roles from this specific role performance: responsible global player, culturally relevant player, and economic power. Her account rests very much on the perspective of mostly decision-makers and remains unclear about the domestic contestation in the larger identity discourse. John also uses role-theoretical language anecdotally but without theoretical rigor and therefore without exhausting the analytical power of the approach. In an analysis of Seoul's middle-power diplomacy, he simply attributes the roles of facilitator, interlocutor, and norm entrepreneur to Korea, without going into greater detail on the formation, contestation, and maintenance of these specific roles.[129]

Hence, it is appropriate to bring a theoretically sound and empirically rich understanding of role theory back into the wealth of identity perspectives, in general, and into the study of South Korean identity, in particular. In his comprehensive account of the history and development of role theory in IR and FPA, Thies highlights the great integrative potential of role theory for both subdisciplines.[130] In this spirit and by going back to the parameters for identity research compiled from Berenskoetter[131] and Stark Urrestarazu,[132] the advantages of taking a role-theoretical perspective to explore agency-oriented identity issues related to foreign policy become apparent.

With the focus on role performance, role theory is very much in line with the performative and situational aspects of identity. For role theorists, any role must be based in a corresponding identity element in order to be meaningful and thus legitimate, and any identity understanding can only be reaffirmed by respective role-playing. A sense of self or a self-identity is thus acquired through the reflection of and interaction with significant and generalized Others. Theoretically, this sense of self is further expressed and habitualized in a narrative way, and either routinely enacted (with the "Me" dominating) or creatively newly calibrated (with the "I" dominating) especially in times of crisis, when old frames of meaningful behavior lose their value. Whereas the former can be tracked by the researcher, the latter puts the analyst into an epistemological impasse. Furthermore, unpacking the identity formation within the "ego" opens the opportunity to also trace the emotional dimension of identification politics along the horizontal and vertical contestation of the formation of role conceptions. As Mercer puts it, "[w]ho we are is what we feel. Identity and emotion depend on each other. Identification without emotion inspires no action for one does not care."[133]

Taking Lebow's criticism about the use of the term *identity* seriously again,[134] it seems only justified to use the term *identity* for the elusive "I" part of the ego. But, likewise, it would make sense to use the term *self-identity* for the observable aspect of the ego, formed through the situational interplay of "I" and "Me," which is narrated for stability and legitimacy. Here, identities are understood as self-descriptions or self-images and are "not related to any particular social function or action in or for a group as such [and t]hey are also unspecific with regard to time and social context."[135] The bearer of the self-identity is here a nation-state as an international foreign policy actor toward other international foreign policy actors or organizations. But the state as a composite actor is itself a social structure in which different political actors, individuals, and political groups challenge the boundaries and direction of the organized collective, the polity. Just to be clear, the government of this state may also engage in similar role-playing with trade unions over labor policies, and there may be an international dimension to this, but this is not what is in focus here.

Because of the situational and performative dimensions of state identity throughout various policy fields, it seems to be either an illusion or at least a very ambitious goal to aim for a complete assessment of the identity of a polity at a given time. Having said that, there are indeed recent positivist attempts to "make identity count,"[136] for example, by building a database of great power national identities from 1810 to 2010 based on "identities" retrieved from official government text sources. Before embarking on such large-scale positivist and government-centric projects, the argument can be made that too many aspects are still unclear and neglected in the study of identity, especially those of smaller or middle-sized countries whose agency and internal sources for identity formation remain outside of the focus of mainstream IR and FPA. The following chapter shows how such an agency-oriented framework for tracing identity can be developed by utilizing role-theoretical concepts.

Notes

1. Abdelal, Rawi, Herrera, Yoshiko M., Johnston, Alastair Iain and McDermott, Rose 2006: "Identity as a Variable", *American Review of Political Science*, Vol. 4, No. 4, p. 695.
2. Brubaker, Rogers and Cooper, Frederick 2000: "Beyond 'Identity'", *Theory and Society*, Vol. 29, No. 1, pp. 1–47.
3. Berenskoetter, Felix 2010: "Identity in International Relations", in: Robert A. Denemark (ed.), *The International Studies Encyclopedia*, Volume 6, Malden, MA: Wiley-Blackwell, p. 3600.
4. Ibid.: p. 3595.
5. See Van Evera, Stephen 1994: "Hypotheses on Nationalism and War", *International Security*, Vol. 18, No. 4, pp. 5–39.

6. See Krasner, Stephen 1999: *Sovereignty: Organized Hypocrisy*, Princeton, NJ: Princeton University Press; Jackson, Robert 2000: *The Global Covenant: Human Conduct in a World of States*, Oxford: Oxford University Press.
7. See Keohane, Robert O. 1989: *International Institutions and State Power: Essays in International Relations Theory*, Boulder, CO: Westview.
8. Harris, Erika 2009: *Nationalism: Theories and Cases*, Edinburgh: Edinburgh University Press, p. 175.
9. Wendt, Alexander 1992: "Anarchy Is What States Make of It: The Social Construction of Power Politics", *International Organization*, Vol. 46, No. 2, pp. 391–425.
10. See also Price, Richard and Reus-Smit, Christian 1998: "Dangerous Liaisons? Critical International Theory and Constructivism", *European Journal of International Relations*, Vol. 4, No. 3, pp. 259–294.
11. Hofferberth, Matthias and Weber, Christian 2015: "Lost in Translation: A Critique of Constructivist Norm Research", *Journal of International Relations and Development*, Vol. 18, No. 1, p. 78.
12. Katzenstein, Peter (ed.) 1996: *The Culture of National Security: Norms and Identity in World Politics*, Cambridge: Cambridge University Press.
13. Wendt, A. 1992; Klotz, Audie 1995: *Norms in International Relations: The Struggle against Apartheid*, Ithaca: Cornell University Press; Finnemore, Martha 1996: *National Interests in International Society*, Ithaca: Cornell University Press.
14. Hofferberth, M. and Weber, C. 2015: p. 78.
15. Campbell, David 1992: *Writing Security: United States Foreign Policy and the Politics of Identity*, Manchester: Manchester University Press; Weldes, Jutta, Laffey, Mark, Gusterson, Hugh and Duvall, Raymond (eds.) 1999: *Cultures of Insecurity: States, Communities, and the Production of Danger*, Minneapolis: University of Minnesota Press.
16. Neumann, Iver B. 1996: *Russia and the Idea of Europe: A Study in Identity and International Relations*, London: Routledge.
17. Wendt, Alexander 1999: *Social Theory of International Politics*, Cambridge: Cambridge University Press.
18. Zehfuss, Maja 2001: "Constructivism and Identity: A Dangerous Liaison", *European Journal of International Relations*, Vol. 7, No. 3, p. 335.
19. Cedermann, Lars Erik and Daase, Christopher 2003: "Endogenizing Corporate Identities: The Next Step in Constructivist IR Theory", *European Journal of International Relations*, Vol. 9, No. 1, pp. 5–35.
20. Ibid.: p. 8.
21. Aggestam, Lisbeth 1999: "Role Conceptions and the Politics of Identity in Foreign Policy", *ARENA Working Paper 8/99*, p. 2. Available at: www.sv.uio.no/arena/english/research/publications/arena-working-papers/1994-2000/1999/wp99_8.htm [Accessed 5 November 2018].
22. McCourt, David M. 2011: "Role-Playing and Identity Affirmation in International Politics: Britain's Reinvasion of the Falklands, 1982", *Review of International Studies*, Vol. 37, No. 4, p. 1606.
23. Legro, Jeffrey 2009: "The Plasticity of Identity under Anarchy", *European Journal of International Relations*, Vol. 15, No. 1, pp. 37–65.
24. Hopf, Ted and Allan, Bentley B. 2016: *Making Identity Count: Building a National Identity Database*, Oxford: Oxford University Press.
25. McCourt, D.M. 2011: p. 1604.
26. Ibid.: p. 1605.
27. McCourt, David M. 2012: "The Roles States Play: A Meadian Interactionist Approach", *Journal of International Relations and Development*, Vol. 15, No. 3, p. 370.

28. See also Checkel, Jeffrey T. 2008: "Constructivism and Foreign Policy", in: Steve Smith, Amelia Hadfield and Tim Dunne (eds.), *Foreign Policy: Theories, Actors, Cases*, Oxford: Oxford University Press, p. 72.
29. Stahl, Bernhard and Harnisch, Sebastian 2009: "Nationale Identitäten und Außenpolitiken: Erkenntnisse, Desiderate und neue Wege in der Diskursforschung" [National Identity and Foreign Policies: Findings, Desiderata and New Paths in Discourse Analysis], in: Bernhard Stahl and Sebastian Harnisch (eds.), *Vergleichende Außenpolitikforschung und nationale Identitäten. Die Europäische Union im Kosovo-Konflikt 1996–2008 [Comparative Foreign Policy Analysis and National Identities: The European Union in the Kosovo Conflict 1996–2008]*, Baden-Baden: Nomos, p. 32.
30. Mead, George H. 1934: *Mind, Self and Society from the Standpoint of a Social Behaviorist*, Chicago: The University of Chicago Press.
31. McCourt, D.M. 2011: p. 1606.
32. Harnisch, Sebastian 2014: "Full-Spectrum Role-Taking: A Two-Level Role Theoretical Role Model", *Conference Paper prepared for Annual Conference of the International Studies Association*, 26–30 March, Toronto, p. 13. Available at: www.uni-heidelberg.de/md/politik/harnisch/person/vortraege/harnisch_2014_-_isa_conference_paper_full_spectrum_role_taking_draft_24_03.pdf [Accessed 5 November 2018].
33. Berenskoetter, F. 2010.
34. Ibid.: p. 3600.
35. Steele, Brent J. 2008: *Ontological Security in International Relations: Self-Identity and the IR State*, London/New York: Routledge; Mitzen, Jennifer 2006: "Ontological Security in World Politics: State Identity and the Security Dilemma", *European Journal of International Relations*, Vol. 12, No. 3, pp. 341–370.
36. Kowert, Paul A. 1999: "National Identity: Inside and Out", in: Glenn Chafetz, Michael Spirtas and Benjamin Frankel (eds.), *The Origins of National Interests*, London: Frank Cass, pp. 1–34; Hopf, Ted 2002: *Social Construction of International Politics*, Ithaca: Cornell University Press.
37. Ringmar, Erik 1996: *Identity, Interest and Action: A Cultural Explanation of Sweden's Intervention in the Thirty Years War*, Cambridge: Cambridge University Press.
38. Hall, Rodney B. 1999: *National Collective Identity: Social Constructs and International Systems*, New York: Columbia University Press, p. 6; Bloom, William 1990: *Personal Identity, National Identity and International Relations*, Cambridge: Cambridge University Press.
39. Berenskoetter, F. 2010: p. 3601.
40. Ibid.
41. Fitzgerald, Thomas K. 1993: *Metaphors of Identity: A Culture-Communication Dialogue*, Albany, NY: State University of New York Press, p. 186.
42. Campbell, D. 1992.
43. Stark Urrestarazu, Ursula 2015: "'Identity' in International Relations and Foreign Policy Theory", in: Knud Erik Jørgensen and Gunther Hellmann (eds.), *Theorizing Foreign Policy in a Globalized World*, London: Palgrave Macmillan, p. 134.
44. Ibid.: p. 135.
45. Ibid.: p. 137.
46. Berenskoetter, Felix 2014: "Parameters of a 'National Biography'", *European Journal of International Relations*, Vol. 20, No. 1, p. 282.
47. Ibid.: p. 269.
48. Ringmar, E. 1996; Weldes, J. et al. 1999; Wæver, Ole 2002: "Identity, Communities and Foreign Policy: Discourse Analysis as Foreign Policy Theory", in: Lene Hansen and Ole Wæver (eds.), *European Integration and National Identity: The*

Challenge of the Nordic States, London/New York: Routledge, pp. 20–49; Bially Mattern, Janice 2005: *Ordering International Politics: Identity, Crisis, and Representational Force*, London/New York: Routledge; Hansen, Lene 2006: *Security as Practice: Discourse Analysis and the Bosnian War*, London/New York: Routledge; Steele, B. 2008.

49. Wehner, Leslie E. and Thies, Cameron G. 2014: "Role Theory, Narratives, and Interpretation: The Domestic Contestation of Roles", *International Studies Review*, Vol. 16, No. 3, pp. 411–436; Wehner, Leslie E. 2018: "The Narration of Roles in Foreign Policy Analysis", *Journal of International Relations and Development*, 25 July 2018, first published online. Available at: https://link.springer.com/article/10.1057/s41268-018-0148-y [Accessed 5 November 2018].
50. Epstein, Liana M., Goff, Philip A., Huo, Yuen J. and Wong, Lauren H. 2013: "A Taste for Justice: Unpacking Identity Politics in a Nascent Democracy", *Political Psychology*, Vol. 34, No. 5, pp. 779–789.
51. Stark Urrestarazu, U. 2015: p. 137.
52. Mead, G.H. 1934.
53. Erikson, Erik 1968: *Identity, Youth and Crisis*, New York: W. W. Norton Company.
54. Berenskoetter, F. 2010: p. 3602.
55. Tajfel, Henri and Turner, John 1986: "The Social Identity Theory of Intergroup Behaviour", in: Stephen Worchel and William G. Austin (eds.), *Psychology of Intergroup Relations*, Chicago: Nelson-Hall Publishers, pp. 7–24.
56. Stark Urrestarazu, U. 2015: p. 133.
57. See also Johnston, Alistair I. 2005: "Conclusions and Extensions: Toward Mid-Range Theorizing and Beyond Europe", *International Organization*, Vol. 59, No. 4, pp. 1013–1044.
58. Adler-Nissen, Rebecca 2016: "The Social Self in International Relations: Identity, Power and the Rediscovery of Constructivism's Symbolic Interactionist Roots", *European Review of International Studies*, Vol. 3, No. 3, pp. 32–33.
59. Welch Larson, Deborah and Shevchenko, Alexei 2010: "Status Seekers Chinese and Russian Responses to U.S. Primacy", *International Security*, Vol. 34, No. 4, pp. 63–95.
60. Lebow, Richard Ned 2008: *A Cultural Theory of International Relations*, Cambridge: Cambridge University Press.
61. Adler-Nissen, R. 2016: p. 33.
62. Emirbayer, Mustafa and Mische, Ann 1998: "What Is Agency?", *The American Journal of Sociology*, Vol. 103, No. 4, pp. 962–1023; Joas, Hans 1996: *The Creativity of Action*, Chicago: University of Chicago Press.
63. Hofferberth, M. and Weber, C. 2015; Bucher, Bernd and Jasper, Ursula 2017: "Revisiting 'Identity' in International Relations: From Identity as Substance to Identifications in Action", *European Journal of International Relations*, Vol. 23, No. 2, pp. 1–26.
64. Hofferberth, M. and Weber, C. 2015: p. 85.
65. Ibid.
66. Ibid.: p. 90.
67. Stark Urrestarazu, U. 2015: p. 139.
68. Lebow, Richard Ned 2016: *National Identifications and International Relations*, Cambridge: Cambridge University Press, p. 79.
69. Ibid.
70. Ibid.: p. 82.
71. McCourt, D.M. 2011: p. 1606; Harnisch, S. 2014: p. 13.
72. See, for example, Crawford, Neta C. 2000: "The Passion of World Politics: Propositions on Emotion and Emotional Relationships", *International Security*, Vol. 24, No. 4, pp. 116–156; Bially Mattern, Janice 2011: "A Practice Theory of Emotion for International Relations", in: Emanuel Adler and Vincent Pouliot (eds.),

International Practices, New York: Cambridge University Press, pp. 63–86; Bleiker, Roland and Hutchison, Emma 2014: "Introduction: Emotions and World Politics", *International Theory*, Vol. 6, No. 3, pp. 490–491; Koschut, Simon 2017: "Introduction to Discourse and Emotions in International Relations", *International Studies Review*, Vol. 19, No. 3, pp. 482–487.

73. Stark Urrestarazu, U. 2015: p. 137.
74. Ibid.
75. Kinnvall, Catarina 2017: "Feeling Ontologically (In)secure: States, Traumas and the Governing of Gendered Space", *Cooperation and Conflict*, Vol. 52, No. 1, pp. 90–108.
76. Mercer, Jonathan 2014: "Feeling Like a State: Social Emotion and Identity", *International Theory*, Vol. 6, No. 3, p. 517.
77. Ibid.: p. 530.
78. Ibid.: p. 524.
79. Forsberg, Tuomas, Heller, Regina and Wolf, Reinhard 2014: "Status and Emotions in Russian Foreign Policy", *Communist and Post-Communist Studies*, Vol. 47, No. 3–4, pp. 261–268; Lindemann, Thomas and Ringmar, Erik (eds.) 2014: *The International Politics of Recognition*, Boulder/London: Paradigm Publishers; Daase, Christopher, Geis, Anna, Fehl, Caroline and Kolliarakis, Georgios (eds.) 2015: *Recognition in International Relations: Rethinking a Political Concept in a Global Context*, London: Palgrave Macmillan.
80. Reus-Smit, Christian 2014: "Emotions and the Social", *International Theory*, Vol. 6, No. 3, p. 571.
81. Ibid.: p. 568.
82. Koschut, S. 2017: p. 484.
83. Berenskoetter, F. 2014: p. 281.
84. Ibid.: p. 283.
85. McCourt, D.M. 2011: p. 1619.
86. Aggestam, Lisbeth 2006: "Role Theory and European Foreign Policy: A Framework of Analysis", in: Ole Elgström and Michael Smith (eds.), *The European Union's Roles in International Politics: Concepts and Analysis*, London/New York: Routledge, p. 12; Breuning, Marijke 2011: "Role Theory Research in International Relations: State of the Art and Blind Spots", in: Sebastian Harnisch, Cornelia Frank and Hans W. Maull (eds.), *Role Theory in International Relations: Approaches and Analyses*, London/New York: Routledge, p. 17.
87. Holsti, Kal J. 1970: "National Role Conceptions in the Study of Foreign Policy", *International Studies Quarterly*, Vol. 13, No. 3, pp. 246–247.
88. Ibid.: p. 239.
89. Harnisch, Sebastian 2018: "Role Theory in International Relations", *Oxford Bibliographies: International Relations*, 11 January. Available at: www.oxfordbibliographies.com/view/document/obo-9780199743292/obo-9780199743292-0226.xml [Accessed 5 November 2018]. For other comprehensive accounts of role theory in International Relations and Foreign Policy Analysis see Breuning, M. 2011 as well as Thies, Cameron G. 2017: "Role Theory and Foreign Policy", in: Robert A. Denemark and Renée Marlin-Bennett (eds.), *The International Studies Encyclopedia*, Malden, MA: Wiley-Blackwell. Available at: www.oxfordreference.com/view/10.1093/acref/9780191842665.001.0001/acref-9780191842665-e-0333?rskey=4M0K6z&result=372#acref-9780191842665-e-0333-div1-0005 [Accessed 5 November 2018].
90. Walker, Stephen G. (ed.) 1987: *Role Theory and Foreign Policy Analysis*, Durham, NC: Duke University Press.
91. Kirste, Knut and Maull, Hanns W. 1996: "Zivilmacht und Rollentheorie" [Civilian Power and Role Theory], *Zeitschrift für Internationale Beziehungen*, Vol. 3, No. 2, pp. 283–312.

92. Barnett, Michael N. 1993: "Institutions, Roles and Disorder: The Case of the Arab States System", *International Studies Quarterly*, Vol. 37, No. 3, pp. 271–296; Cronin, Bruce 2001: "The Paradox of Hegemony: America's Ambiguous Relationship with the United Nations", *European Journal of International Relations*, Vol. 7, No. 1, pp. 103–130.
93. Thies, Cameron G. 2013b: *The United States, Israel and the Search for International Order: Socializing States*, New York: Routledge.
94. Cantir, Cristian and Kaarbo, Juliet (eds.) 2016: *Domestic Role Contestation, Foreign Policy and International Relations*, London/New York: Routledge.
95. Breuning, M. 2011: p. 20.
96. Wendt, A. 1992.
97. Thies, Cameron G. 2013a: "The Roles of Bipolarity: A Role Theoretic Understanding of the Effects of Ideas and Material Factors on the Cold War", *International Studies Perspectives*, Vol. 14, No. 3, pp. 269–288.
98. Thies, C.G. 2013b.
99. Nabers, Dirk 2011: "Identity and Role Change in International Politics", in: Sebastian Harnisch, Cornelia Frank and Hans W. Maull (eds.), *Role Theory in International Relations: Approaches and Analyses*, London/New York: Routledge, p. 75.
100. Adler-Nissen, R. 2016: p. 36.
101. Harnisch, Sebastian, Frank, Cornelia and Maull, Hans W. 2011: "Introduction", in: Sebastian Harnisch, Cornelia Frank and Hans W. Maull Hans (eds.), *Role Theory in International Relations: Approaches and Analyses*, London/New York: Routledge, p. 2.
102. Sandstrom, Kent, Martin, Daniel and Fine, Gary 2010: *Symbols, Selves, and Social Reality: A Symbolic Interactionist Approach to Social Psychology*, Oxford: Oxford University Press.
103. For example, Harnisch, Sebastian, Frank, Cornelia and Maull, Hans W. (eds.) 2011: *Role Theory in International Relations: Approaches and Analyses*, London/New York: Routledge; McCourt, D.M. 2011, 2012; Harnisch, Sebastian 2012: "Conceptualizing in the Minefield: Role Theory and Foreign Policy Learning", *Foreign Policy Analysis*, Vol. 8, No. 1, pp. 47–69; Harnisch, Sebastian, Bersick, Sebastian and Gottwald, Jörn-Carsten (eds.) 2016: *China's International Roles*, London/New York: Routledge; Cantir, Cristian and Kaarbo, Juliet 2012: "Contested Roles and Domestic Politics: Reflections on Role Theory in Foreign Policy Analysis and IR Theory", *Foreign Policy Analysis*, Vol. 8, No. 1, pp. 5–24; Thies, C.G. 2013a, 2013b; Wehner, L.E. and Thies, C.G. 2014; Brummer, Klaus and Thies, Cameron G. 2015: "The Contested Selection of National Role Conceptions", *Foreign Policy Analysis*, Vol. 11, No. 3, pp. 273–293; Cantir, C. and Kaarbo, J. 2016.
104. Mead, G.H. 1934: p. 154.
105. Harnisch, Sebastian 2016a: "Role Theory and the Study of Chinese Foreign Policy", in: Sebastian Harnisch, Sebastian Bersick and Jörn-Carsten Gottwald (eds.), *China's International Roles*, London/New York: Routledge, p. 10.
106. Ibid.: p. 11.
107. Mead, G.H. 1934: p. 178.
108. Ibid.: p. 174.
109. Harnisch, S. 2012: p. 53.
110. Beneš, Vít and Harnisch, Sebastian 2015: "Role Theory in Symbolic Interactionism: Czech Republic, Germany and the EU", *Cooperation and Conflict*, Vol. 50, No. 1, pp. 146–165.
111. Harnisch, Sebastian 2016b: "China's Historical Self and Its International Role", in: Sebastian Harnisch, Sebastian Bersick and Jörn-Carsten Gottwald (eds.), *China's International Roles*, London/New York: Routledge, pp. 38–58.
112. Harnisch, S. 2016a: p. 8.

113. Ibid.: p. 11.
114. Burke, Peter J. and Stets, Jan E. 2009: *Identity Theory*, Oxford: Oxford University Press, p. 202.
115. Harnisch, S. 2016a: p. 8.
116. Ibid.: pp. 8–9.
117. Cantir, C. and Kaarbo, J. 2016.
118. Nabers, D. 2011.
119. Kaarbo, Juliet and Cantir, Cristian 2013: "Role Conflict in Recent Wars: Danish and Dutch Debates over Iraq and Afghanistan", *Cooperation and Conflict*, Vol. 48, No. 4, pp. 467–468.
120. Cantir, C. and Kaarbo, J. 2016: p. 18.
121. Flockhart, Trine 2006: "Complex Socialization: A Framework for the Study of State Socialization", *European Journal of International Relations*, Vol. 12, No. 1, p. 92; McCourt, D.M. 2011: p. 1619.
122. Nabers, D. 2011: p. 84.
123. Ibid.
124. Compare Lebow, Richard Ned 2012: *The Politics and Ethics of Identity: In Search of Ourselves*, Cambridge: Cambridge University Press.
125. Checkel, Jeffrey T. 2005: "International Institutions and Socialization in Europe: Introduction and Framework", *International Organization*, Vol. 59, No. 4, pp. 801–826.
126. Easley, Leif-Eric 2012: "Middle Power National Identity? South Korea and Vietnam in US-China Geopolitics", *Pacific Focus*, Vol. 27, No. 3, pp. 421–442.
127. Kim, Sungjin 2014: "South Korea's Climate Diplomacy: Analysis Based on the Perspective of 'Middle Power Diplomacy'", *EAI Middle Power Diplomacy Initiative Working Paper*, No. 5, Seoul: East Asia Institute, p. 3.
128. Hermanns, Heike 2013: "National Role Conceptions in the 'Global Korea' Foreign Policy Strategy", *The Korean Journal of International Studies*, Vol. 11, No. 1, pp. 55–82.
129. John, Jojin V. 2014: "Becoming and Being a Middle Power: Exploring a New Dimension of South Korea's Foreign Policy", *China Report*, Vol. 50, No. 4, pp. 325–341.
130. Thies, C.G. 2017.
131. Berenskoetter, F. 2010, 2014.
132. Stark Urrestarazu, U. 2015.
133. Mercer, J. 2014: p. 522.
134. Lebow, R.N. 2016.
135. Harnisch, S. 2014: p. 13.
136. Hopf, T. and Allan, B.B. 2016.

References

Abdelal, Rawi, Herrera, Yoshiko M., Johnston, Alastair Iain and McDermott, Rose 2006: "Identity as a Variable", *American Review of Political Science*, Vol. 4, No. 4, pp. 695–711.

Adler-Nissen, Rebecca 2016: "The Social Self in International Relations: Identity, Power and the Rediscovery of Constructivism's Symbolic Interactionist Roots", *European Review of International Studies*, Vol. 3, No. 3, pp. 27–39.

Aggestam, Lisbeth 1999: "Role Conceptions and the Politics of Identity in Foreign Policy", *ARENA Working Paper 8/99*. Available at: www.sv.uio.no/arena/english/research/publications/arena-working-papers/1994-2000/1999/wp99_8.htm [Accessed 5 November 2018].

Aggestam, Lisbeth 2006: "Role Theory and European Foreign Policy: A Framework of Analysis", in: Ole Elgström and Michael Smith (eds.), *The European Union's Roles in International Politics: Concepts and Analysis*, London/New York: Routledge, pp. 11–29.

Barnett, Michael N. 1993: "Institutions, Roles and Disorder: The Case of the Arab States System", *International Studies Quarterly*, Vol. 37, No. 3, pp. 271–296.

Beneš, Vít and Harnisch, Sebastian 2015: "Role Theory in Symbolic Interactionism: Czech Republic, Germany and the EU", *Cooperation and Conflict*, Vol. 50, No. 1, pp. 146–165.

Berenskoetter, Felix 2010: "Identity in International Relations", in: Robert A. Denemark (ed.), *The International Studies Encyclopedia*, Volume 6, Malden, MA: Wiley-Blackwell, pp. 3595–3611.

Berenskoetter, Felix 2014: "Parameters of a 'National Biography'", *European Journal of International Relations*, Vol. 20, No. 1, pp. 262–288.

Bially Mattern, Janice 2005: *Ordering International Politics: Identity, Crisis, and Representational Force*, London/New York: Routledge.

Bially Mattern, Janice 2011: "A Practice Theory of Emotion for International Relations", in: Emanuel Adler and Vincent Pouliot (eds.), *International Practices*, New York: Cambridge University Press, pp. 63–86.

Bleiker, Roland and Hutchison, Emma 2014: "Introduction: Emotions and World Politics", *International Theory*, Vol. 6, No. 3, pp. 490–491.

Bloom, William 1990: *Personal Identity, National Identity and International Relations*, Cambridge: Cambridge University Press.

Breuning, Marijke 2011: "Role Theory Research in International Relations: State of the Art and Blind Spots", in: Sebastian Harnisch, Cornelia Frank and Hans W. Maull (eds.), *Role Theory in International Relations: Approaches and Analyses*, London/New York: Routledge, pp. 16–35.

Brubaker, Rogers and Cooper, Frederick 2000: "Beyond 'Identity'", *Theory and Society*, Vol. 29, No. 1, pp. 1–47.

Brummer, Klaus and Thies, Cameron G. 2015: "The Contested Selection of National Role Conceptions", *Foreign Policy Analysis*, Vol. 11, No. 3, pp. 273–293.

Bucher, Bernd and Jasper, Ursula 2017: "Revisiting 'Identity' in International Relations: From Identity as Substance to Identifications in Action", *European Journal of International Relations*, Vol. 23, No. 2, pp. 1–26.

Burke, Peter J. and Stets, Jan E. 2009: *Identity Theory*, Oxford: Oxford University Press.

Campbell, David 1992: *Writing Security: United States Foreign Policy and the Politics of Identity*, Manchester: Manchester University Press.

Cantir, Cristian and Kaarbo, Juliet 2012: "Contested Roles and Domestic Politics: Reflections on Role Theory in Foreign Policy Analysis and IR Theory", *Foreign Policy Analysis*, Vol. 8, No. 1, pp. 5–24.

Cantir, Cristian and Kaarbo, Juliet (eds.) 2016: *Domestic Role Contestation, Foreign Policy and International Relations*, London/New York: Routledge.

Cedermann, Lars Erik and Daase, Christopher 2003: "Endogenizing Corporate Identities: The Next Step in Constructivist IR Theory", *European Journal of International Relations*, Vol. 9, No. 1, pp. 5–35.

Checkel, Jeffrey T. 2005: "International Institutions and Socialization in Europe: Introduction and Framework", *International Organization*, Vol. 59, No. 4, pp. 801–826.

Checkel, Jeffrey T. 2008: "Constructivism and Foreign Policy", in: Steve Smith, Amelia Hadfield and Tim Dunne (eds.), *Foreign Policy: Theories, Actors, Cases*, Oxford: Oxford University Press, pp. 71–82.

Crawford, Neta C. 2000: "The Passion of World Politics: Propositions on Emotion and Emotional Relationships", *International Security*, Vol. 24, No. 4, pp. 116–156.

Cronin, Bruce 2001: "The Paradox of Hegemony: America's Ambiguous Relationship with the United Nations", *European Journal of International Relations*, Vol. 7, No. 1, pp. 103–130.

Daase, Christopher, Geis, Anna, Fehl, Caroline and Kolliarakis, Georgios (eds.) 2015: *Recognition in International Relations: Rethinking a Political Concept in a Global Context*, London: Palgrave Macmillan.

Easley, Leif-Eric 2012: "Middle Power National Identity? South Korea and Vietnam in US-China Geopolitics", *Pacific Focus*, Vol. 27, No. 3, pp. 421–442.

Emirbayer, Mustafa and Mische, Ann 1998: "What Is Agency?", *The American Journal of Sociology*, Vol. 103, No. 4, pp. 962–1023.

Epstein, Liana M., Goff, Philip A., Huo, Yuen J. and Wong, Lauren H. 2013: "A Taste for Justice: Unpacking Identity Politics in a Nascent Democracy", *Political Psychology*, Vol. 34, No. 5, pp. 779–789.

Erikson, Erik 1968: *Identity, Youth and Crisis*, New York: W. W. Norton Company.

Finnemore, Martha 1996: *National Interests in International Society*, Ithaca: Cornell University Press.

Fitzgerald, Thomas K. 1993: *Metaphors of Identity: A Culture-Communication Dialogue*, Albany, NY: State University of New York Press.

Flockhart, Trine 2006: "Complex Socialization: A Framework for the Study of State Socialization", *European Journal of International Relations*, Vol. 12, No. 1, pp. 89–118.

Forsberg, Tuomas, Heller, Regina and Wolf, Reinhard 2014: "Status and Emotions in Russian Foreign Policy", *Communist and Post-Communist Studies*, Vol. 47, No. 3–4, pp. 261–268.

Hall, Rodney B. 1999: *National Collective Identity: Social Constructs and International Systems*, New York: Columbia University Press.

Hansen, Lene 2006: *Security as Practice: Discourse Analysis and the Bosnian War*, London/New York: Routledge.

Harnisch, Sebastian 2012: "Conceptualizing in the Minefield: Role Theory and Foreign Policy Learning", *Foreign Policy Analysis*, Vol. 8, No. 1, pp. 47–69.

Harnisch, Sebastian 2014: "Full-Spectrum Role-Taking: A Two-Level Role Theoretical Role Model", *Conference Paper prepared for Annual Conference of the International Studies Association*, 26–30 March, Toronto. Available at: www.uni-heidelberg.de/md/politik/harnisch/person/vortraege/harnisch_2014_-_isa_conference_paper_full_spectrum_role_taking_draft_24_03.pdf [Accessed 5 November 2018].

Harnisch, Sebastian 2016a: "Role Theory and the Study of Chinese Foreign Policy", in: Sebastian Harnisch, Sebastian Bersick and Jörn-Carsten Gottwald (eds.), *China's International Roles*, London/New York: Routledge, pp. 3–21.

Harnisch, Sebastian 2016b: "China's Historical Self and Its International Role", in: Sebastian Harnisch, Sebastian Bersick and Jörn-Carsten Gottwald (eds.), *China's International Roles*, London/New York: Routledge, pp. 38–58.

Harnisch, Sebastian 2018: "Role Theory in International Relations", *Oxford Bibliographies: International Relations*, 11 January. Available at: www.oxfordbibliographies.

com/view/document/obo-9780199743292/obo-9780199743292-0226.xml [Accessed 5 November 2018].

Harnisch, Sebastian, Bersick, Sebastian and Gottwald, Jörn-Carsten (eds.) 2016: *China's International Roles*, London/New York: Routledge.

Harnisch, Sebastian, Frank, Cornelia and Maull, Hans W. (eds.) 2011a: *Role Theory in International Relations: Approaches and Analyses*, London/New York: Routledge.

Harnisch, Sebastian, Frank, Cornelia and Maull, Hans W. 2011b: "Introduction", in: Sebastian Harnisch, Cornelia Frank and Hans W. Maull (eds.), *Role Theory in International Relations: Approaches and Analyses*, London/New York: Routledge, pp. 1–4.

Harris, Erika 2009: *Nationalism: Theories and Cases*, Edinburgh: Edinburgh University Press.

Hermanns, Heike 2013: "National Role Conceptions in the 'Global Korea' Foreign Policy Strategy", *The Korean Journal of International Studies*, Vol. 11, No. 1, pp. 55–82.

Hofferberth, Matthias and Weber, Christian 2015: "Lost in Translation: A Critique of Constructivist Norm Research", *Journal of International Relations and Development*, Vol. 18, No. 1, pp. 75–103.

Holsti, Kal J. 1970: "National Role Conceptions in the Study of Foreign Policy", *International Studies Quarterly*, Vol. 13, No. 3, pp. 233–309.

Hopf, Ted 2002: *Social Construction of International Politics*, Ithaca: Cornell University Press.

Hopf, Ted and Allan, Bentley B. 2016: *Making Identity Count: Building a National Identity Database*, Oxford: Oxford University Press.

Jackson, Robert 2000: *The Global Covenant: Human Conduct in a World of States*, Oxford: Oxford University Press.

Joas, Hans 1996: *The Creativity of Action*, Chicago: University of Chicago Press.

John, Jojin V. 2014: "Becoming and Being a Middle Power: Exploring a New Dimension of South Korea's Foreign Policy", *China Report*, Vol. 50, No. 4, pp. 325–341.

Johnston, Alistair I. 2005: "Conclusions and Extensions: Toward Mid-Range Theorizing and Beyond Europe", *International Organization*, Vol. 59, No. 4, pp. 1013–1044.

Kaarbo, Juliet and Cantir, Cristian 2013: "Role Conflict in Recent Wars: Danish and Dutch Debates over Iraq and Afghanistan", *Cooperation and Conflict*, Vol. 48, No. 4, pp. 465–483.

Katzenstein, Peter (ed.) 1996: *The Culture of National Security: Norms and Identity in World Politics*, Cambridge: Cambridge University Press.

Keohane, Robert O. 1989: *International Institutions and State Power: Essays in International Relations Theory*, Boulder, CO: Westview.

Kim, Sungjin 2014: "South Korea's Climate Diplomacy: Analysis Based on the Perspective of 'Middle Power Diplomacy'", *EAI Middle Power Diplomacy Initiative Working Paper*, No. 5, Seoul: East Asia Institute.

Kinnvall, Catarina 2017: "Feeling Ontologically (In)secure: States, Traumas and the Governing of Gendered Space", *Cooperation and Conflict*, Vol. 52, No. 1, pp. 90–108.

Kirste, Knut and Maull, Hanns W. 1996: "Zivilmacht und Rollentheorie" [Civilian Power and Role Theory], *Zeitschrift für Internationale Beziehungen*, Vol. 3, No. 2, pp. 283–312.

Klotz, Audie 1995: *Norms in International Relations: The Struggle against Apartheid*, Ithaca: Cornell University Press.

Koschut, Simon 2017: "Introduction to Discourse and Emotions in International Relations", *International Studies Review*, Vol. 19, No. 3, pp. 482–487.

Kowert, Paul A. 1999: "National Identity: Inside and Out", in: Glenn Chafetz, Michael Spirtas and Benjamin Frankel (eds.), *The Origins of National Interests*, London: Frank Cass, pp. 1–34.

Krasner, Stephen 1999: *Sovereignty: Organized Hypocrisy*, Princeton, NJ: Princeton University Press.

Lebow, Richard Ned 2008: *A Cultural Theory of International Relations*, Cambridge: Cambridge University Press.

Lebow, Richard Ned 2012: *The Politics and Ethics of Identity: In Search of Ourselves*, Cambridge: Cambridge University Press.

Lebow, Richard Ned 2016: *National Identifications and International Relations*, Cambridge: Cambridge University Press.

Legro, Jeffrey 2009: "The Plasticity of Identity under Anarchy", *European Journal of International Relations*, Vol. 15, No. 1, pp. 37–65.

Lindemann, Thomas and Ringmar, Erik (eds.) 2014: *The International Politics of Recognition*, Boulder/London: Paradigm Publishers.

McCourt, David M. 2011: "Role-Playing and Identity Affirmation in International Politics: Britain's Reinvasion of the Falklands, 1982", *Review of International Studies*, Vol. 37, No. 4, pp. 1599–1621.

McCourt, David M. 2012: "The Roles States Play: A Meadian Interactionist Approach", *Journal of International Relations and Development*, Vol. 15, No. 3, pp. 370–392.

Mead, George H. 1934: *Mind, Self and Society from the Standpoint of a Social Behaviorist*, Chicago: The University of Chicago Press.

Mercer, Jonathan 2014: "Feeling Like a State: Social Emotion and Identity", *International Theory*, Vol. 6, No. 3, pp. 515–535.

Mitzen, Jennifer 2006: "Ontological Security in World Politics: State Identity and the Security Dilemma", *European Journal of International Relations*, Vol. 12, No. 3, pp. 341–370.

Nabers, Dirk 2011: "Identity and Role Change in International Politics", in: Sebastian Harnisch, Cornelia Frank and Hans W. Maull Hans (eds.), *Role Theory in International Relations: Approaches and Analyses*, London/New York: Routledge, pp. 74–92.

Neumann, Iver B. 1996: *Russia and the Idea of Europe: A Study in Identity and International Relations*, London: Routledge.

Price, Richard and Reus-Smit, Christian 1998: "Dangerous Liaisons? Critical International Theory and Constructivism", *European Journal of International Relations*, Vol. 4, No. 3, pp. 259–294.

Reus-Smit, Christian 2014: "Emotions and the Social", *International Theory*, Vol. 6, No. 3, pp. 568–574.

Ringmar, Erik 1996: *Identity, Interest and Action: A Cultural Explanation of Sweden's Intervention in the Thirty Years War*, Cambridge: Cambridge University Press.

Sandstrom, Kent, Martin, Daniel and Fine, Gary 2010: *Symbols, Selves, and Social Reality: A Symbolic Interactionist Approach to Social Psychology*, Oxford: Oxford University Press.

Stahl, Bernhard and Harnisch, Sebastian 2009: "Nationale Identitäten und Außenpolitiken: Erkenntnisse, Desiderate und neue Wege in der Diskursforschung" [National Identity and Foreign Policies: Findings, Desiderata and New Paths in Discourse Analysis], in: Bernhard Stahl and Sebastian Harnisch (eds.), *Vergleichende Außenpolitikforschung und nationale Identitäten. Die Europäische Union im Kosovo-Konflikt 1996–2008 [Comparative Foreign Policy Analysis and national identities: The European Union in the Kosovo Conflict 1996–2008]*, Baden-Baden: Nomos, pp. 31–58.

Stark Urrestarazu, Ursula 2015: "'Identity' in International Relations and Foreign Policy Theory", in: Knud Erik Jørgensen and Gunther Hellmann (eds.), *Theorizing Foreign Policy in a Globalized World*, London: Palgrave Macmillan, pp. 126–149.

Steele, Brent J. 2008: *Ontological Security in International Relations: Self-Identity and the IR State*, London/New York: Routledge.

Tajfel, Henri and Turner, John 1986: "The Social Identity Theory of Intergroup Behaviour", in: Stephen Worchel and William G. Austin (eds.), *Psychology of Intergroup Relations*, Chicago: Nelson-Hall Publishers, pp. 7–24.

Thies, Cameron G. 2013a: "The Roles of Bipolarity: A Role Theoretic Understanding of the Effects of Ideas and Material Factors on the Cold War", *International Studies Perspectives*, Vol. 14, No. 3, pp. 269–288.

Thies, Cameron G. 2013b: *The United States, Israel and the Search for International Order: Socializing States*, New York: Routledge.

Thies, Cameron G. 2017: "Role Theory and Foreign Policy", in: Robert A. Denemark and Renée Marlin-Bennett (eds.), *The International Studies Encyclopedia*, Malden, MA: Wiley-Blackwell. Available at: www.oxfordreference.com/view/10.1093/acref/9780191842665.001.0001/acref-9780191842665-e-0333?rskey=4M0K6z&result=372#acref-9780191842665-e-0333-div1-0005 [Accessed 5 November 2018].

Van Evera, Stephen 1994: "Hypotheses on Nationalism and War", *International Security*, Vol. 18, No. 4, pp. 5–39.

Wæver, Ole 2002: "Identity, Communities and Foreign Policy: Discourse Analysis as Foreign Policy Theory", in: Lene Hansen and Ole Wæver (eds.), *European Integration and National Identity: The Challenge of the Nordic States*, London/New York: Routledge, pp. 20–49.

Walker, Stephen G. (ed.) 1987: *Role Theory and Foreign Policy Analysis*, Durham, NC: Duke University Press.

Wehner, Leslie E. 2018: "The Narration of Roles in Foreign Policy Analysis", *Journal of International Relations and Development*, 25 July 2018, first published online. Available at: https://link.springer.com/article/10.1057/s41268-018-0148-y [Accessed 5 November 2018].

Wehner, Leslie E. and Thies, Cameron G. 2014: "Role Theory, Narratives, and Interpretation: The Domestic Contestation of Roles", *International Studies Review*, Vol. 16, No. 3, pp. 411–436.

Welch Larson, Deborah and Shevchenko, Alexei 2010: "Status Seekers Chinese and Russian Responses to U.S. Primacy", *International Security*, Vol. 34, No. 4, pp. 63–95.

Weldes, Jutta, Laffey, Mark, Gusterson, Hugh and Duvall, Raymond (eds.) 1999: *Cultures of Insecurity: States, Communities, and the Production of Danger*, Minneapolis: University of Minnesota Press.

Wendt, Alexander 1992: "Anarchy Is What States Make of It: The Social Construction of Power Politics", *International Organization*, Vol. 46, No. 2, pp. 391–425.

Wendt, Alexander 1999: *Social Theory of International Politics*, Cambridge: Cambridge University Press.

Zehfuss, Maja 2001: "Constructivism and Identity: A Dangerous Liaison", *European Journal of International Relations*, Vol. 7, No. 3, pp. 315–348.

4 A Model for Inquiry into International State Identity

The aim of this study is an inquiry into the international identity of the nation-state ROK: How does South Korea relate itself to the world in the early 21st century, or put differently, what kind of international identity is underpinning the ROK's rise on the global stage? The earlier demonstrated empirical gap with regard to South Korean identity and global foreign policy led to the conclusion that South Korea's global foreign policy is potentially linked to different identity narratives, from nationalism and postcolonialism over pan-Asianism to globalism. In addition, the literature on South Korean identity led to some skepticism, especially with regard to the alleged influence of a "globalist" or liberal internationalist self-identity in the South Korean international identity discourse. As has been further argued in this study, by conflating role and identity, in treating the state as a unitary actor and by solely categorizing actors according to their behavior (as, for example, a middle power), we remain unable to understand South Korea's own take on its place in the world, and thus also Seoul's international behavior. Such a simplistic understanding of identity is obstructing our view regarding the actual identity dynamics that would allow us, first, to understand the ROK better as a foreign policy actor and, second, to grant South Koreans some agency in the shaping of their relationship with the world.

In line with the parameters earlier identified for research into state identity, in this book the nation-state is not treated as a unitary actor but as a social structure itself within which different domestic political actors compete for dominance. In terms of methodology, this study grasps the international identity of the state actor ROK through focusing on practices of self-identification in two selected foreign policy fields, international peacekeeping and climate diplomacy. Self-identification practices are here defined as specific links between role conceptions and self-images by different political actors within the domestic setting of the ROK. Through an inductive, textual analysis of these respective foreign policy discourses, one can account for the different self-identifications at play in different foreign policy decisions, while the focus lies on the domestic contestation and stability of these self-understandings over time. The comparison of these retrieved self-images in these self-identifications from both policy fields and over time allows inference of larger patterns of international self-identity narratives. As snapshots of an international, collective sense of self, these self-images enable insights into the larger international identity dynamics behind Korea's global foreign policy.

Practices of Self-Identification in Foreign Policy Discourses

As one can see from the review of literature on identity from IR constructivism in the previous chapter, identity is not easily conceptualized for empirical applications, and thus identity studies must be clear and upfront about the level of analysis, the location of agency, and the political actors considered to be acting on behalf of the state in order to plausibly and reproducibly account for the politics of international state identity (re-)formation. Here, an approach based on symbolic interactionist role-theoretical concepts is presented to theorize the link between foreign policy and state identity and to empirically investigate the international identity of the ROK in global governance. Due to this role-theoretical approach, this study provides an agency-oriented account of the ROK's international state identity in selected foreign policy fields while also doing justice to the fragility and complexity of the South Korean state's self-identity on the global stage.[1]

This interpretative approach can be situated in FPA and the IR field of International Political Sociology that work with "a relational and processual ontology which endogenizes interest or identity formation, resists different forms of reifications of structures, institutions or agents/subjects, as well as materialist or idealist reductionism."[2] As such, the approach developed here stands—like most IR constructivism—in the tradition of Max Weber's interpretative sociology that focuses on *verstehen* or *understanding* of social action from the perspective of agents and how this assigned meaning to and made sense of their actions and the world around them.[3] For IR theory, Hollis and Smith term this postpositivist approach of understanding an insider's story, "told so as to make us understand what the events mean, in a sense distinct from any meaning found in unearthing the laws of nature,"[4] and Wendt calls it the question of "how are things in the world put together so that they have the properties that they do?"[5] Accordingly, the main research question at hand asks not so much about the causality behind but the constitution of South Korea's international agency or, in other words, the relationship between its emergence as a key global player and South Korean state identity.

The symbolic interactionist role-theoretical identity perspective at hand emphasizes the agency of actors as not simply norm dupes or zombie-like executors of a basic character but as active interpreters of social attitudes. Moreover, as McCourt puts it, "roles are the necessary social vehicle for action in its meaning-creating, identity-affirming sense."[6] Hence, roles are understood as emergent relational outcomes of an ongoing process of interaction between the Self and significant and generalized Others, whereas the contestation over role conceptions within a composite actor is about the legitimation of actions in reference to certain identity elements. These identity elements or self-images are consequently reaffirmed through role performance. In consequence, self-identity here is not fixed but constantly formed through practices of self-identification or the linking of self-images and role conceptions to make sense of and justify certain political actions. In this conceptualization, South Korea has one state identity only, even though it is contested from within through different self-identifications.[7] In line with Zehfuss's notion, an actor's identity here is more than just its social manifestation in role identities and thus more than "something negotiable between states."[8] Importantly, however, because of their situational and performative character, state actors' identities can only be grasped through these social manifestations.

In this particular Korean case, where there is another state that claims to be Korean, Pyeongyang's Other-identification from the outside may theoretically also be relevant for self-identification within South Korea but only if domestic political actors choose to tap into self-images originating from the North. Practically, however, with the National Security Law restricting any pro-North political agitation, this seems less likely. Furthermore, it is often simply a state actor's role-playing and thus role identity that changes with a change of government but not the more fundamental identity. It may help to think of this as differently utilized or activated identity elements or self-images that have been part of the political discourse all along, which are then linked to role conceptions by domestic political actors, forming role identities of the state once enacted. Also, role-theoretical works have often been focused on the emergence of roles and stability of orders, whereas in this study identity is investigated through the utilization of role-theoretical concepts. This approach leads us to two observable layers of the international self-identity of the state actor ROK: the layer of concretely manifested role identities in a given social context and the more basic layer of self-images that have been employed for the legitimation of specific role conceptions and related actions. Finally, going beyond the layer of a state's role identities in the analysis enables us to gain insights into or snapshots of the broader international identity of South Korea.

Accordingly, this study proposes to start the inquiry into the ROK's international sense of self from actual role performance or foreign policy actions, because this angle allows accounting for the performative and situational aspects of identity. As all foreign policy actions must be based in an identity in order to be meaningful, and because identity can only be actualized and manifested through performance,[9] the suggestion here is to start an inquiry into the self-identity of a nation-state actor by analyzing the contested decision-making process about role conceptions in concrete action-demanding situations. The formation of role conceptions in the decision-making discourse is where self-images are utilized in order to legitimate actions and are consequently reaffirmed through role performance.[10] Hence, this interpretative approach not only is in line with narrative analyses in role theory[11] and perspectives that emphasize identification practices[12] but also gels with the relational-processual research program that suggests looking "at the interpretations, reasons and justifications that agents offer for their actions and to investigate how they change."[13]

A crucial focus lies here on the domestic horizontal contestation of role conception, where political elites have to justify their role conceptions (and role behavior) and win the support of other political actors and the citizenry in accordance with larger identity narratives.[14] By interpretatively reconstructing these domestic debates about role conceptions over a longer period, during different events and over different policy fields, one will be able to map how certain role conceptions relate to specific self-images as practices of self-identification. One can further think of these distinct self-identifications as different international selves that, in all their variety, constitute the international state identity, in contrast to socially manifest international role identities and the broader international identity of the actor (see Figure 4.1).

Importantly, foreign policy practice cannot simply be inferred from identity elements retrieved independently from the practice in question, because every action depends on an actual mobilization of identity elements by actors in any given situation. The same actor can mobilize the same self-image in different social contexts to a different outcome, because these identity elements are undetermined and can be linked to different role conceptions. It is this link between evasive identity elements and contested role conceptions that lends legitimacy to specific role performances as practices of self-identification that serve as the entry point for inquiries into international identity.

This is illustrated in Figure 4.1, where, for example, the same self-image, here self-image b, was linked by discursive agents to two different role

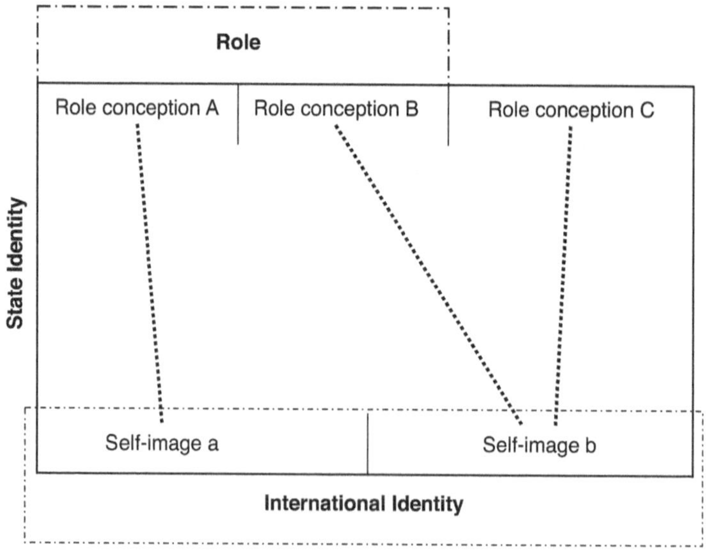

Figure 4.1 Role, State Identity, and International Identity—Practices of Self-identification in Foreign Policy Discourses

Source: Created by author.

conceptions, B and C. However, only role conception B ultimately supported a socially recognized role. An example of such a case would be the German decision to participate in the 1998–1999 Kosovo War. On one hand, the German self-image of being responsible for the Holocaust was used to justify military action against another feared genocide in the Balkans, whereas on the other hand, the German self-image of being responsible for World War II and the Holocaust led domestic political actors to put forward a role conception of Germany as a nonmilitary, civilian power. As we know, this struggle over the right role for Germany in this context ultimately led to the country's participation in NATO–led intervention. In this exemplary case we have two distinct patterns of self-identification practices both referring to the same self-image: first, Germany as a humanitarian-interventionist power because of the narrative of historical responsibility and, second, Germany as a nonmilitary, civilian power because of the narrative of historical responsibility. Again, in symbolic interactionist role theory, it is through the interpretative process of self-identification that the Self is constituted.

At this point it is also important to clearly differentiate again the terms *role*, *role identity*, and *role conception*, as these terms are not to be used interchangeably. A role is a temporary social position determined by the interaction between the Self and Others with a corresponding behavioral pattern. A role in this sense is a structural category and can be enacted by different actors, often leading to counter-roles, such as "leader" and "follower." In a concrete social relationship, however, when an actor has successfully established (domestically/the Self-identification and internationally/Other-identification) and enacted the role of, for example, "responsible member," one can speak of role identity. The perspective here is that of the actor in contrast with other more established, systemic perspectives focusing on norms, for example.

Every socially manifested role identity and enacted role is domestically established through discursive practices of self-identification, which are here defined as linkages between role conceptions and self-images. In the foreign policy discourse, however, it is possible that different role conceptions together support an actually established role identity. This is the case where there are different narratives between government parties, for example, in coalition governments, where a minor role conception supports a specific foreign policy action and thus the dominant role conception as well as the related role enactment. Moreover, deviant role conceptions from the political opposition that counter the official government role conceptions are not role identities from this perspective, as they were not successfully established domestically or internationally and, as a consequence, were not enacted.

Furthermore, a self-image is a description of "oneself" throughout time and space and a narrative that situates the actor narratively in the past and present. Self-images are in this sense clear identity articulations, answers to the question, "Who are we?" These identity elements are utilized as legitimation tools for specific actions in specific social contexts. For example, a discursive actor may articulate South Korea's colonial history to argue against overseas troop deployments as a "fighter for postcolonial justice" or a "defender of the sovereignty norm." This would initially constitute role conceptions as ideas in a political debate and potentially role identities, once these have been successfully established and enacted. The historical memory of colonization, however, could also be utilized to argue for military action and overseas troop deployments, because such an intervention was aimed at restoring the sovereignty of another postcolonial state against "imperialist aggression." Here we would have a role conception that entails a certain leadership aspiration, because of the perceived duty or right to interfere in the affairs of other sovereign states in order to preserve or restore sovereignty. It becomes clear from the preceding example that a role conception is a socially contextual idea about how to relate to an Other with clear, directed actions that are purposeful, because the conception builds on respective self-images, which, in turn, lend it meaning. Again, self-images are generally undetermined and can be linked to different role conceptions. It is the linking between self-images as evasive identity elements and contested role conceptions that lend legitimacy to specific role performances as practices of self-identification that serve as the entry point for inquiries into international state identity in relation to foreign policy.

For empirical analyses, the first task lies in identifying key events of foreign policy decisions relevant for a particular role enactment as starting points for in-depth analysis. The second step follows an inductive, textual analysis to retrieve the various role conceptions and related goals and proposed actions that have been employed in the legitimation of the selected foreign policy decisions. As there is no universally agreed catalogue about roles in international relations,[15] an inductive close reading of the text seems to be more appropriate than a deductive content analysis.[16] Finally, after tracing the self-images as part of the role conceptions referenced in the foreign policy discourse, an intertextual, interpretative reading of all these self-identifications, understood as links between self-image and role conception, will lead as "snapshots" to larger identity narratives. The argument here is that only the comprehensive mapping of these different self-identifications (on both layers of role identities/conceptions and self-images) deserves to be called an acceptable assessment of a state's international identity, because such a description is reflecting their contestation and stability over different foreign policy fields and over time. In addition, the retrieved self-images provide snapshots of the broader international identity of South Korea.

The guiding questions for the textual analysis of self-identifications in specific discourses of foreign policy decisions are the following:

1. What were the potential roles considered by which discursive agents?
2. Which options/actions were seen as appropriate and recommended?
3. What are the related goals?
4. How was a specific role conception legitimated?
5. Which identity representations (self-images) have been utilized?
6. How successful were those attempts and were they contested?

These questions are inspired by Stahl and Harnisch[17] and are based on George and Bennett's "method of structured, focused comparison"[18] for case study approaches. This set of general questions is structured in the sense that they are standardized and applied to all cases in order to enable meaningful comparison. Furthermore, the method is focused because it is applied with a specific theoretically informed research objective. Thus, this approach will enable the analyst to retrieve the specific practices of self-identification understood as distinct links between role conceptions and self-images. The comparison of the different self-identifications from different foreign policy fields then allows insights into larger self-identity narratives at play. The degree of contestation over self-identifications and the stability of these patterns of self-identification over time are the main indicators for the prevalence and stability in the content of these wider self-identification practices. Only if there are indeed hegemonic and stable patterns of self-identification over time and over different foreign policy fields is it justified to speak of one of these as the "international identity" of the ROK. If this is not the case, then the complexity and contestation of self-identifications in these foreign policy fields must be adequately reflected in any account of "international identity" of a state actor.

Furthermore, the agency-oriented approach opens the black box of the nation-state actor and focuses on horizontal domestic role contestation, which "disaggregates the ego into a number of actors that contest how the state should behave."[19] Generally, new self-images and role conceptions must be more or less in line with or adapt to established self-understandings: "New ideas about social order and the nation state need to resonate with previously embedded and institutionalized values, symbols and myths."[20] But one should also keep in mind that political agents are very likely choosing only ideas that are in line with or supporting their own interests in what Marcussen et al. call "interest assumption."[21] As soon as a self-understanding becomes consensual, "the number of legitimate ideas available in a political discourse decreases and the institutionalization of ideas in institutions and political culture makes them resistant to challenges."[22] Obviously, not all political actors are equally influential in this struggle over the right role behavior of the nation-state. Depending on the respective decision-making processes, economic, military, political, institutional, or charismatic authority will be distributed differently between political actors. The rather transparent democratic decision-making processes are also easier to observe and analyze than those in totalitarian regimes, for example, in North Korea. Just because it is hard to assess the different power factions behind the ambiguity and secrecy of Pyeongyang politics, it does not mean that there is no contestation over the "right" role behavior of the state as well.

In the case of South Korea, the relevant political actors in the "wider foreign policy debate"[23] are the government, consisting first and foremost of the president, the ministers, and ministry officials, as well as the political opposition. In the Republic's presidential system, the president, as commander-in-chief and head of state, possesses considerable authority over foreign affairs and the legislative process and is elected by popular vote for a single five-year term. In practice, presidents often try to shape their personal legacies by pursuing their own big and bold strategies and initiatives with little regard for their predecessor's work, which ultimately hinders consensus and continuity, especially over foreign policy issues.[24] The National Assembly, the South Korean unicameral parliament, is one of the key venues for horizontal role contestation[25] between elites, because it is here that the government has to build support for its actions and justify its intended role behavior toward the political opposition and the public. Furthermore, in the South Korean political system, both the president and the parliament have the right to initiate legislation.[26] Because Rathbun points to the role of political parties as "vehicles for ideologies"[27] that "bring ideas with them through the front door,"[28] Cantir and Kaarbo note that "[p]arliaments allow for representation of a wide range of viewpoints and thus may be a forum for discourse on and contestation of national role conceptions."[29]

The South Korean political landscape is dominated by two ideologically distinct party blocs, the conservatives (*bosu*) and the progressives (*jinbo*), with several small fringe parties on the right as well as the left. Typical for South Korean party politics is the lack of institutionalization and relative fluidity[30] that result in regular rebrandings of the personalized party platforms, mostly due to changes in leadership and less so because of contentious changes of policy or ideology.[31] The two-bloc structure has been very stable since the 1990s with the, as of 2018, progressive-left Democratic or Minjoo Party of Korea and the conservative Liberty Korea Party being the main umbrella outfits for various splinter factions within their ideological blocs. The defining ideological cleavage is the foreign policy and national security question of how to pursue relations with North Korea and the US, respectively.[32] Whereas the Minjoo Party has been supporting engagement with Pyeongyang and a more autonomous stance on the US alliance is forming the so-called self-reliance faction (*jajupa*), the conservatives forming the so-called alliance faction (*dongmaengpa*) have been traditionally in strong support of a close US alliance and a more hardline approach to North Korea.[33] Apart from these politically contentious issues, however, there has been traditionally a rather broad consensus on most other foreign policy issues in the National Assembly.[34]

An inquiry into the self-identification practices of state actors and their foreign policies is focused on the narratives of legitimation in the political decision-making process and does not include, for example, media discourse or public opinions. Future researchers may want to include these additional sources for a more detailed and comprehensive depiction of a greater diversity of self-identifications, but if these more marginal or deviant self-understandings from the media or public sphere are to have some political effect, they will have to be fed into the political decision-making process at some point. With the constitutive approach suggested here, one is mainly able to grasp the state identity formation related to foreign policy in retrospect. If, however, a study is primarily interested in prognoses or likelihoods of future foreign policy and identity developments, it is advisable to include media and public discourses in the research design in order to also account for self-identifications that have not yet been politicized by actors in the political system. Therefore, the state identity–focused methodological approach and scope of this study can only serve as pixelated snapshots of the broad South Korean international identity. But, as a starting point, they can illustrate where and how to focus the analytical lens beyond state identity for further similar inquiries into, for example, media, civil society, or cultural discourses.

Methodology

The goal of the study at hand is to investigate precisely which self-identifications relate to state identity and in consequence which snapshots of international identity can be observed in the legitimation of various global foreign policy actions in the domestic foreign policy discourse. South Korea's rise in global affairs takes place in various foreign policy fields. For example, Seoul is active in global trade, contributes to international peace and security, has emerged as a major donor in international developmental aid, and takes part in global human-rights debates. The country is playing a leading role in international climate diplomacy and pursues scientific and political interests through its space program[35] and polar strategy.[36] Any complete account of South Korean international state identity in global affairs must cover all these policy areas in what would amount to a large-scale research task for a whole research group.

Given the limited scale of this study, focusing on only a selection of policy fields seems legitimate. Thus, the empirical analysis of the relationship between South Korean international state identity and Seoul's global foreign policy is turning to two salient foreign policy fields of South Korea's global outreach as identified by the related literature: South Korean international peacekeeping[37] and Seoul's climate diplomacy.[38] These two fields are fitting for analysis as they are not only central for national security, sovereignty, and prosperity but also for multilateralism and supranational governance. Both selected policy fields are also at the intersection of other relevant foreign policy matters: whereas international peacekeeping as an issue touches upon the US–ROK security alliance and intra-Korean relations, climate diplomacy is closely related to issues about South Korea's international trade competitiveness and developmental assistance. Furthermore, Seoul showed the greatest leadership aspirations in those two policy fields and, if a somehow "globalist" self-identification has indeed started to develop in the South Korean international identity discourse, as is one subquestion of this study, it must be observable here.

The outstanding period of South Korea's global outreach was with no doubt the presidency of Lee Myung-bak, a conservative, former Hyundai CEO, and former mayor of Seoul, from 2008 to 2013. Under his leadership, the government promoted the vision of "Global Korea" as a national security strategy aimed at building "a nation that aims to contribute to global peace and development under a broader vision and a more proactive approach to interacting with the international community."[39] Key issue areas in the strategy were free trade, developmental assistance, peacekeeping, climate change, and energy matters as well as cultural diplomacy. Seoul also made several successful efforts to host major international gatherings, such as the 2010 G20 summit, the HLF-4, which took place in Busan in 2011, and the second NSS in 2012. It is with regard to this diplomatic focus on global leadership and agenda setting that observers[40] agree to attribute to South Korea the status of an "important middle-ranking power in global affairs."[41] John argues from a post-structural perspective that "Global Korea was not merely a point of departure in Korean foreign policy but was also the key site of Korean national identity construction."[42]

Because of this, the presidency of Lee Myung-bak is at the center of this study at hand, but the observation period for the empirical study is extended to include the previous presidency of the progressive Roh Moo-hyun from 2003 to 2008 and the succeeding presidency of the conservative Park Geun-hye from 2013 to late 2016, just before her impeachment. In doing so, the whole observation period of 14 years, from 2003 to 2016, covers conservative and progressive governments and different parliamentary majorities, as well as a before and after of the formative events of the Global Korea vision. As indicated earlier, the study also looks at two different policy fields, peacekeeping and climate diplomacy, in order to allow for a cross-sectoral and cross-ideological comparison of the self-identifications retrieved from the empirical analysis.

The first case study on international peacekeeping focuses on key decisions made in the South Korean National Assembly, such as the troop deployments to the second United Nations Interim Force in Lebanon (UNIFIL II) in 2006, the disaster relief-focused UN Stabilization Mission in Haiti (MINUSTAH) from 2010 until 2012, and the deployment to the UN Mission in South Sudan (UNMISS) since 2013. A key event in this policy area was the debate over the legislation of the "United Nations Peacekeeping Operations Participation Act" in 2009, which has been the legal framework for peacekeeping deployments ever since. With regard to the second case study on international climate diplomacy, the key events identified are the parliamentary debate about the "Framework Act on Low Carbon and Green Growth" in 2009, the establishment of the Global Green Growth Institute in 2012, and the successful bid to host the United Nations Green Climate Fund in Songdo, Incheon City, in 2012, as well as the ratification of the Paris Agreement in 2016.

Sources for the textual analysis are all statements made by government representatives, from the president to ministry officials, as well as the political opposition and parties. These sources were available in digitalized form, mostly in the online archives of speeches, policy, and press statements of government officials on the respective websites of these institutions,[43] or sometimes interviews or opinion pieces in newspapers or think-tank publications. Most of these were, in addition to the Korean original, available in an English-language version. After taking a random sample of sources, a check of the English-language translation provided by the respective institutions confirmed that there were no substantive discrepancies between the different language versions. The parliamentary debate from plenary, standing committee, and special committee to subcommittee sessions is accessible through the digital archive of the National Assembly Minutes[44] in text and video, both in Korean language only. The respective sources have been translated by the author, who bears the sole responsibility for any mistakes in this regard.

The organization, inductive coding, and interpretation of the generated data collection were conducted manually by the author with the help of the qualitative data analysis computer software package NVivo, produced by QSR International. A list of the nodes for the retrieved self-images from the NVivo database can be found in the Appendix for illustration. All in all, the analysis yielded results from 2005 to 2016 in the form of approximately 150 different source materials with retrieved identity elements, such as speeches, statements, and parliamentary recordings, sometimes including more than one speaker.[45]

Last, in an effort to address the issue of ethical consequences of this study, it should be acknowledged that it is, in this case, a non-Korean, nonnative speaker of Korean and researcher outside of the country who is contributing to an academic representation of what "Korea" *is* in the world. The approach followed here is quite literally "taking the role of Korea," and one should be aware of the limitations of this positionality, as well as of the responsibilities. With a study like this, Korea is cast and framed in a specific way from the outside in academic discourse, while the challenge to negotiate between the expectations evoked by this representation and "self-identifications" is left to Korea and Koreans. Having said that, one goal of this approach at hand is to shift the focus from mostly external categorization of South Korean behavior to a perspective that tries to take South Korean agency seriously. This can, of course, only be a starting point for a different approach for studying international state identity in the case of South Korea. In terms of research beneficence, outsiderness, however, may have the potential benefit of offering a less or at least differently obstructed view on the issues of identity and agency at hand. As such, it can only ever be a starting or complementary take on what South Korean identity *is* in the 21st century.

Notes

1. Compare Adler-Nissen, Rebecca 2016: "The Social Self in International Relations: Identity, Power and the Rediscovery of Constructivism's Symbolic Interactionist Roots", *European Review of International Studies*, Vol. 3, No. 3, p. 32.
2. Guzzini, Stefano 2016: "International Political Sociology, or: The Social Ontology and Power Politics of Process", *DIIS Working Paper 06*, p. 5. Available at: http://pure.diis.dk/ws/files/556451/DIIS_WP_2016_6.pdf [Accessed 5 November 2018].
3. Weber, Max 1968: *Economy and Society: An Outline of Interpretive Sociology*, New York: Bedminster Press.
4. Hollis, Martin and Smith, Steve 1990: *Explaining and Understanding International Relations*, Oxford: Clarendon Press, p. 1.
5. Wendt, Alexander 1998: "On Constitution and Causation in International Relations", *Review of International Studies*, Vol. 24, p. 103.

6. McCourt, David M. 2012: "The Roles States Play: A Meadian Interactionist Approach", *Journal of International Relations and Development*, Vol. 15, No. 3, p. 370.
7. Compare McCourt, David M. 2011: "Role-Playing and Identity Affirmation in International Politics: Britain's Reinvasion of the Falklands, 1982", *Review of International Studies*, Vol. 37, No. 4, p. 1606; Harnisch, Sebastian 2014: "Full-Spectrum Role-Taking: A Two-Level Role Theoretical Role Model", *Conference Paper prepared for Annual Conference of the International Studies Association*, 26–30 March, Toronto, p. 13. Available at: www.uni-heidelberg.de/md/politik/harnisch/person/vortraege/harnisch_2014_-_isa_conference_paper_full_spectrum_role_taking_draft_24_03.pdf [Accessed 5 November 2018].
8. Zehfuss, Maja 2001: "Constructivism and Identity: A Dangerous Liaison", *European Journal of International Relations*, Vol. 7, No. 3, p. 335.
9. McCourt, D.M. 2011: p. 1600.
10. Flockhart, Trine 2006: "Complex Socialization: A Framework for the Study of State Socialization", *European Journal of International Relations*, Vol. 12, No. 1, p. 92; McCourt, D.M. 2011: p. 1619.
11. Wehner, Leslie E. and Thies, Cameron G. 2014: "Role Theory, Narratives, and Interpretation: The Domestic Contestation of Roles", *International Studies Review*, Vol. 16, No. 3, pp. 411–436; Wehner, Leslie E. 2018: "The Narration of Roles in Foreign Policy Analysis", *Journal of International Relations and Development*, 25 July 2018, first published online. Available at: https://link.springer.com/article/10.1057/s41268-018-0148-y [Accessed 5 November 2018].
12. Lebow, Richard Ned 2016: *National Identifications and International Relations*, Cambridge: Cambridge University Press.
13. Hofferberth, Matthias and Weber, Christian 2015: "Lost in Translation: A Critique of Constructivist Norm Research", *Journal of International Relations and Development*, Vol. 18, No. 1, p. 90.
14. Harnisch, Sebastian 2012: "Conceptualizing in the Minefield: Role Theory and Foreign Policy Learning", *Foreign Policy Analysis*, Vol. 8, No. 1, p. 63; see also Abdelal, Rawi, Herrera, Yoshiko M., Johnston, Alastair Iain and McDermott, Rose 2006: "Identity as a Variable", *American Review of Political Science*, Vol. 4, No. 4, pp. 695–711.
15. For an early compilation of international roles, see Holsti, Kal J. 1970: "National Role Conceptions in the Study of Foreign Policy", *International Studies Quarterly*, Vol. 13, No. 3, pp. 255, 296.
16. Cantir, Cristian and Kaarbo, Juliet 2016: "Unpacking Ego in Role Theory: Vertical and Horizontal Role Contestation and Foreign Policy", in: Cristian Cantir and Juliett Kaarbo (eds.), *Domestic Role Contestation, Foreign Policy and International Relations*, London/New York: Routledge, p. 19.
17. Stahl, Bernhard and Harnisch, Sebastian 2009: "Nationale Identitäten und Außenpolitiken: Erkenntnisse, Desiderate und neue Wege in der Diskursforschung" [National Identity and Foreign Policies: Findings, Desiderata and New Paths in Discourse Analysis], in: Bernhard Stahl and Sebastian Harnisch (eds.), *Vergleichende Außenpolitikforschung und nationale Identitäten. Die Europäische Union im Kosovo-Konflikt 1996–2008 [Comparative Foreign Policy Analysis and National Identities: The European Union in the Kosovo Conflict 1996–2008]*, Baden-Baden: Nomos, p. 51.
18. George, Alexander L. and Bennett, Andrew 2005: *Case Studies and Theory Development in the Social Sciences*, Cambridge, MA: MIT Press, p. 67ff.
19. Cantir, C. and Kaarbo, J. 2016: p. 6.
20. Marcussen, Martin, Risse, Thomas, Engelmann-Martin, Daniela, Knopf, Hans Joachim and Roscher, Klaus 1999: "Constructing Europe? The Evolution of French, British and German Nation State Identities", *Journal of European Public Policy*, Vol. 6, No. 4, p. 616.

21. Ibid.: p. 617.
22. Ibid.
23. Hansen, Lene 2006: *Security as Practice: Discourse Analysis and the Bosnian War*, London/New York: Routledge, p. 57.
24. Kim, Sung-Mi 2016: "South Korea's Middle-Power Diplomacy: Changes and Challenges", *Research Paper, Chatham House Asia Programme*, 22 June. Available at: www.chathamhouse.org/sites/files/chathamhouse/publications/research/2016-06-22-south-korea-middle-power-kim.pdf [Accessed 5 November 2018].
25. Cantir, C. and Kaarbo, J. 2016: p. 12.
26. Hix, Simon and Jun, Hae-won 2009: "Party Behaviour in the Parliamentary Arena: The Case of the Korean National Assembly", *Party Politics*, Vol. 15, No. 6, p. 668.
27. Rathbun, Brian C. 2004: *Partisan Interventions: European Party Politics and Peace Enforcement in the Balkans*, Ithaca/London: Cornell University Press, p. 7.
28. Ibid.
29. Cantir, C. and Kaarbo, J. 2016: p. 12.
30. Hoon, Jaung 2010: "Political Parties and Democracy in South Korea", in: Baogang He, Anatoly Kulik and Kay Lawson (eds.), *Political Parties and Democracy, Volume II: Post-Soviet and Asian Political Parties*, Santa Barbara, CA: Praeger Publishers, p. 239.
31. See also Park, Cheol Hee 2012: "South Korea", in: Takashi Inoguchi and Jean Blandel (eds.), *Political Parties and Democracy, Contemporary Western Europe and Asia*, New York: Palgrave Macmillan, pp. 127–142.
32. Park, C.H. 2012: pp. 137–138; Hix, S. and Jun, H.W. 2009: p. 689.
33. Chae, Haesook and Kim, Steven 2008: "Conservatives and Progressives in South Korea", *The Washington Quarterly*, Vol. 31, No. 4, pp. 77–95; Kim, S.M. 2016: p. 4.
34. Hix, S. and Jun, H.W. 2009: pp. 689–690; see also Chae, H.S. and Kim, S. 2008.
35. Kang, Tae-jun 2015: "South Korea's Quest to Be a Major Space Power", *The Diplomat*, 27 March. Available at: http://thediplomat.com/2015/03/south-koreas-quest-to-be-a-major-space-power/ [Accessed 5 November 2018].
36. Brady, Anne-Marie and Kim, Seungryeol 2013: "Cool Korea: Korea's Growing Antarctic Interests", in: Anne-Marie Brady (ed.), *The Emerging Politics of Antarctica*, London/New York: Routledge, pp. 75–95; Watson, Iain 2016: "From Middle Power to Pivot Power: Korea as an Arctic Observer in the Age of Eurasia", *Pacific Focus*, Vol. 31, No. 3, pp. 333–356.
37. Hwang, Balbina Y. 2012: "Korea and PKO: Is Korea Contributing to Global Peace?", in: Scott A. Snyder (ed.), *Global Korea: South Korea's Contributions to International Security*, New York/Washington, DC: Council on Foreign Relations, pp. 13–27; Ikenberry, G. John and Mo, Jongryn 2013: *The Rise of Korean Leadership, Emerging Powers and International Order*, New York: Palgrave Macmillan, p. 145; Heo, Uk and Roehrig, Terence 2014: *South Korea's Rise: Economic Development, Power, and Foreign Relations*, Cambridge: Cambridge University Press, p. 172.
38. Saxer, Carl J. 2013: "Capabilities and Aspirations: South Korea's Rise as a Middle Power", *Asia Europe Journal*, Vol. 11, No. 4, pp. 397–413; Ikenberry, G.J. and Mo, J.R. 2013.
39 *Cheong Wa Dae* 2009: "Global Korea: The National Security Strategy of the Republic of Korea: The Lee Myung-bak Administration's Foreign Policy and National Security Vision", June, p. 3.
40. Nye, Joseph S. 2009: "South Korea's Growing Soft Power", *Project Syndicate*, 10 November. Available at: www.project-syndicate.org/commentary/south-korea-s-growing-soft-power?barrier=accessreg [Accessed 5 November 2018]; Ikenberry, G.J. and Mo, J.R. 2013; Saxer, C.J. 2013; Kim, Sangbae 2014: "Roles of Middle Power in East Asia: A Korean Perspective", *East Asia Institute MPDI*

Working Paper, No. 2, 29 January. Available at: www.eai.or.kr/data/bbs/eng_report/20140203158563.pdf [Accessed 5 November 2018].

41. Nye, J. 2009.
42. John, Jojin V. 2015: "Globalization, National Identity and Foreign Policy: Understanding 'Global Korea'", *The Copenhagen Journal of Asian Studies*, Vol. 33, No. 2, p. 38.
43. The URLs for these online sources are for President Roh Moo-hyun: http://16cwd.pa.go.kr/cwd/en/archive/archive_list.php?meta_id=en_speeches&navi=president. President Lee Myung-bak's webpage is archived under the following URL: http://17cwd.pa.go.kr/kr/index.php. President Park Geun-hye's archived webpage seems to be currently (November 2018) only accessible through the Presidential Archive's main webpage: http://pa.go.kr/portal/com/viewMainPage.do. The webpage of the incumbent President, currently Moon Jae-in, is the *Cheong Wa Dae* webpage: http://english1.president.go.kr. For the Ministry of Foreign Affairs: www.mofa.go.kr/ENG/press/speeches/minister/incumbent/index.jsp?menu=m_10_40_10; for the Ministry of National Defense: www.mnd.go.kr/mbshome/mbs/mnd_eng/; and for the Ministry of Environment: http://eng.me.go.kr/eng/web/main.do.
44. The URL for accessing the National Assembly Minutes search interface is http://likms.assembly.go.kr/record/mhs-60-010.do#none.
45. See the Appendix.

References

Abdelal, Rawi, Herrera, Yoshiko M., Johnston, Alastair Iain and McDermott, Rose 2006: "Identity as a Variable", *American Review of Political Science*, Vol. 4, No. 4, pp. 695–711.

Adler-Nissen, Rebecca 2016: "The Social Self in International Relations: Identity, Power and the Rediscovery of Constructivism's Symbolic Interactionist Roots", *European Review of International Studies*, Vol. 3, No. 3, pp. 27–39.

Brady, Anne-Marie and Kim, Seungryeol 2013: "Cool Korea: Korea's Growing Antarctic Interests", in: Anne-Marie Brady (ed.), *The Emerging Politics of Antarctica*, London/New York: Routledge, pp. 75–95.

Cantir, Cristian and Kaarbo, Juliet 2016: "Unpacking Ego in Role Theory: Vertical and Horizontal Role Contestation and Foreign Policy", in: Cristian Cantir and Juliett Kaarbo (eds.), *Domestic Role Contestation, Foreign Policy and International Relations*, London/New York: Routledge, pp. 1–22.

Chae, Haesook and Kim, Steven 2008: "Conservatives and Progressives in South Korea", *The Washington Quarterly*, Vol. 31, No. 4, pp. 77–95.

Cheong Wa Dae 2009: "Global Korea: The National Security Strategy of the Republic of Korea: The Lee Myung-bak Administration's Foreign Policy and National Security Vision", June.

Flockhart, Trine 2006: "Complex Socialization: A Framework for the Study of State Socialization", *European Journal of International Relations*, Vol. 12, No. 1, pp. 89–118.

George, Alexander L. and Bennett, Andrew 2005: *Case Studies and Theory Development in the Social Sciences*, Cambridge, MA: MIT Press.

Guzzini, Stefano 2016: "International Political Sociology, or: The Social Ontology and Power Politics of Process", *DIIS Working Paper 06*. Available at: http://pure.diis.dk/ws/files/556451/DIIS_WP_2016_6.pdf [Accessed 5 November 2018].

Hansen, Lene 2006: *Security as Practice: Discourse Analysis and the Bosnian War*, London/New York: Routledge.

Harnisch, Sebastian 2012: "Conceptualizing in the Minefield: Role Theory and Foreign Policy Learning", *Foreign Policy Analysis*, Vol. 8, No. 1, pp. 47–69.

Harnisch, Sebastian 2014: "Full-Spectrum Role-Taking: A Two-Level Role Theoretical Role Model", *Conference Paper prepared for Annual Conference of the International Studies Association*, 26–30 March, Toronto. Available at: www.uni-heidelberg.de/md/politik/harnisch/person/vortraege/harnisch_2014_-_isa_conference_paper_full_spectrum_role_taking_draft_24_03.pdf [Accessed 5 November 2018].

Heo, Uk and Roehrig, Terence 2014: *South Korea's Rise: Economic Development, Power, and Foreign Relations*, Cambridge: Cambridge University Press.

Hix, Simon and Jun, Hae-won 2009: "Party Behaviour in the Parliamentary Arena: The Case of the Korean National Assembly", *Party Politics*, Vol. 15, No. 6, pp. 667–694.

Hofferberth, Matthias and Weber, Christian 2015: "Lost in Translation: A Critique of Constructivist Norm Research", *Journal of International Relations and Development*, Vol. 18, No. 1, pp. 75–103.

Hollis, Martin and Smith, Steve 1990: *Explaining and Understanding International Relations*, Oxford: Clarendon Press.

Holsti, Kal J. 1970: "National Role Conceptions in the Study of Foreign Policy", *International Studies Quarterly*, Vol. 13, No. 3, pp. 233–309.

Hoon, Jaung 2010: "Political Parties and Democracy in South Korea", in: Baogang He, Anatoly Kulik and Kay Lawson (eds.), *Political Parties and Democracy, Volume II: Post-Soviet and Asian Political Parties*, Santa Barbara, CA: Praeger Publishers, pp. 221–241.

Hwang, Balbina Y. 2012: "Korea and PKO: Is Korea Contributing to Global Peace?", in: Scott A. Snyder (ed.), *Global Korea: South Korea's Contributions to International Security*, New York/Washington, DC: Council on Foreign Relations, pp. 13–27.

Ikenberry, G. John and Mo, Jongryn 2013: *The Rise of Korean Leadership, Emerging Powers and International Order*, New York: Palgrave Macmillan.

John, Jojin V. 2015: "Globalization, National Identity and Foreign Policy: Understanding 'Global Korea'", *The Copenhagen Journal of Asian Studies*, Vol. 33, No. 2, pp. 38–57.

Kang, Tae-jun 2015: "South Korea's Quest to Be a Major Space Power", *The Diplomat*, 27 March. Available at: http://thediplomat.com/2015/03/south-koreas-quest-to-be-a-major-space-power/ [Accessed 5 November 2018].

Kim, Sangbae 2014: "Roles of Middle Power in East Asia: A Korean Perspective", *East Asia Institute MPDI Working Paper*, No. 2, 29 January. Available at: www.eai.or.kr/data/bbs/eng_report/20140203158563.pdf [Accessed 5 November 2018].

Kim, Sung-Mi 2016: "South Korea's Middle-Power Diplomacy: Changes and Challenges", *Research Paper, Chatham House Asia Programme*, 22 June. Available at: www.chathamhouse.org/sites/files/chathamhouse/publications/research/2016-06-22-south-korea-middle-power-kim.pdf [Accessed 5 November 2018].

Lebow, Richard Ned 2016: *National Identifications and International Relations*, Cambridge: Cambridge University Press.

Marcussen, Martin, Risse, Thomas, Engelmann-Martin, Daniela, Knopf, Hans Joachim and Roscher, Klaus 1999: "Constructing Europe? The Evolution of French, British and German Nation State Identities", *Journal of European Public Policy*, Vol. 6, No. 4, pp. 614–633.

McCourt, David M. 2011: "Role-Playing and Identity Affirmation in International Politics: Britain's Reinvasion of the Falklands, 1982", *Review of International Studies*, Vol. 37, No. 4, pp. 1599–1621.

McCourt, David M. 2012: "The Roles States Play: A Meadian Interactionist Approach", *Journal of International Relations and Development*, Vol. 15, No. 3, pp. 370–392.

Nye, Joseph S. 2009: "South Korea's Growing Soft Power", *Project Syndicate*, 10 November. Available at: www.project-syndicate.org/commentary/south-korea-s-growing-soft-power?barrier=accessreg [Accessed 5 November 2018].

Park, Cheol Hee 2012: "South Korea", in: Takashi Inoguchi and Jean Blandel (eds.), *Political Parties and Democracy, Contemporary Western Europe and Asia*, New York: Palgrave Macmillan, pp. 127–142.

Rathbun, Brian C. 2004: *Partisan Interventions: European Party Politics and Peace Enforcement in the Balkans*, Ithaca/London: Cornell University Press.

Saxer, Carl J. 2013: "Capabilities and Aspirations: South Korea's Rise as a Middle Power", *Asia Europe Journal*, Vol. 11, No. 4, pp. 397–413.

Stahl, Bernhard and Harnisch, Sebastian 2009: "Nationale Identitäten und Außenpolitiken: Erkenntnisse, Desiderate und neue Wege in der Diskursforschung" [National Identity and Foreign Policies: Findings, Desiderata and New Paths in Discourse Analysis], in: Bernhard Stahl and Sebastian Harnisch (eds.), *Vergleichende Außenpolitikforschung und nationale Identitäten. Die Europäische Union im Kosovo-Konflikt 1996–2008 [Comparative Foreign Policy Analysis and National Identities: The European Union in the Kosovo Conflict 1996–2008]*, Baden-Baden: Nomos, pp. 31–58.

Watson, Iain 2016: "From Middle Power to Pivot Power: Korea as an Arctic Observer in the Age of Eurasia", *Pacific Focus*, Vol. 31, No. 3, pp. 333–356.

Weber, Max 1968: *Economy and Society: An Outline of Interpretive Sociology*, New York: Bedminster Press.

Wehner, Leslie E. 2018: "The Narration of Roles in Foreign Policy Analysis", *Journal of International Relations and Development*, 25 July 2018, first published online. Available at: https://link.springer.com/article/10.1057/s41268-018-0148-y [Accessed 5 November 2018].

Wehner, Leslie E. and Thies, Cameron G. 2014: "Role Theory, Narratives, and Interpretation: The Domestic Contestation of Roles", *International Studies Review*, Vol. 16, No. 3, pp. 411–436.

Wendt, Alexander 1998: "On Constitution and Causation in International Relations", *Review of International Studies*, Vol. 24, pp. 101–117.

Zehfuss, Maja 2001: "Constructivism and Identity: A Dangerous Liaison", *European Journal of International Relations*, Vol. 7, No. 3, pp. 315–348.

5 South Korea and Peacekeeping

The first case study on self-identification practices in the South Korean global foreign policy discourse is focused on the policy field of UN PKOs, which has been identified as one of the key areas of South Korean global foreign policy ambition in the respective literature.[1] First, the history of, as well as the academic literature on, South Korean peacekeeping is briefly examined, followed by the results of the empirical analysis. The focus hereby lies on role conceptions and key themes exhibited and contested by different actors in the South Korean peacekeeping discourse of the last decade. Finally, the retrieved self-images, as they are linked to role conceptions as specific practices of self-identification, allows a small glimpse of the larger identity narratives underlying South Korea's global foreign policy. The main finding of this case study is that the main role conception is of the hegemonic "responsible member of the international community," linked to a self-image of a risen South Korea, a once peripheral and marginal country that has entered the ranks of advanced countries and is eager to contribute to the "international community" in order to acquire self-assurance through external recognition. However, the duty or responsibility to participate in UN PKOs is not only linked discursively to the reassurance of status but also the historical experience of being helped by the UN during and after the Korean War.

An Introduction to South Korean Peacekeeping

Shortly after finally becoming a UN member state in 1991 as a "relative latecomer,"[2] South Korea's first contribution to UN peacekeeping was in 1993, when Seoul sent an engineering battalion of 252 soldiers to Somalia as part of UNOSOM II.[3] During the 1990s, mostly engineering units, medics, or military observers were deployed in nonmilitary assignments to places like Western Sahara, Bosnia, Angola, Georgia, India, and Pakistan. Only in 1999 did President Kim Dae-jung decide for the first time to send combat forces, an infantry battalion, as a contribution to the International Force for East Timor (INTERFET), a multinational force (MNF) led by Australia that subsequently turned into a PKO under a UN mandate.

Since then, South Korea's involvement with UN peacekeeping has grown increasingly and in 2009 the National Assembly passed the "United Nations

Peacekeeping Operations Participation Act" in order to speed up and streamline the procedure for future deployments under the UN umbrella.[4] Other measures that were part of the PKO Act are the establishment of a standing unit for overseas deployment, the Onnuri Unit (meaning "the whole world"), and the expansion of the Peacekeeping Training Center at the National Defense University.[5] Recently, South Korean peacekeepers have served in more robust assignments, such as the 350-personnel strong Dongmyeong Unit as part of the UNIFIL II since 2007, the disaster relief–focused 240-person engineering-focused Danbi Unit to the MINUSTAH from 2010 until 2012, and the deployment of the Hanbit Unit with just under 300 medics, engineers, and infantry to the UNMISS since 2013. Today, Seoul has deployed around 120,000 personnel to various assignments in 128 countries, with approximately 642 personnel being deployed at the end of 2016.[6]

Because of the geopolitical situation on the Korean Peninsula, military deployment decisions have always been a politically sensitive matter. For a long time, overseas troop deployments, whether in PKO or MNF assignments, were mostly seen either as a necessity of or an opportunity to strengthen the US–ROK defense alliance. Hwang, for example, refers to PKOs as a traditional "utilitarian tool of [US] alliance management" with little independent value for South Korean foreign policy makers.[7] Given the tense security environment on the peninsula, Seoul is said to "privilege [. . .] commitments to the US over commitments to the UN"[8] when having to decide on the deployment of ultimately limited defense capabilities. From a military standpoint, however, PKOs have been seen as a welcome opportunity to gain operational experience in an international environment.[9]

In terms of public support for PKOs abroad, the literature on South Korean peacekeeping identifies a widespread sentiment that South Korea is a "child of the UN" and hence has a moral obligation to "pay back" the international community.[10] This development from being "the host of the largest UN enforcement operation to date"[11] in the Korean War to being a significant troop contributor country nowadays is a widespread narrative regularly shared by South Korean decision-makers. Furthermore, in his assessment of South Korean overseas troop deployments, Hong contrasts this "paying back syndrome" with a "Vietnam syndrome," stemming from the contentious South Korean participation in the Vietnam War.[12] In order to comply with requests from the US, its main ally and geopolitical protector, and in exchange for Washington's investment commitments, hefty compensations and, most important, lucrative business opportunities in the US war economy, South Korea deployed about 300,000 soldiers to Vietnam until 1973.[13] According to an Asan Institute survey from 2012, 57 percent of Koreans believed the participation in the Vietnam War was the right choice and around half, 54 percent, "think that ROK forces were deployed to gain economic benefits for Korea."[14] According to Breuker, in mainstream historiography today the Vietnam War is mostly seen as a way of "jump-starting" South Korea's export-driven economic development under the military dictatorship of Park Chung-hee.[15] Still, Hong sees a key challenge in how PKOs are legitimated domestically.[16] Especially for the political left, the participation in the Vietnam War "was a wrong decision [. . .] forced by the United States."[17]

Even despite their general support for PKOs, the main concern of the political parties in the National Assembly, as well as the nation's population, is the safety of the deployed personnel. Understandably, then, Ruffa notes in her comparative study of different armies in the UNIFIL II mission that the Korean army, being rather new to peacekeeping, chose a comparatively cautious operational approach with a perceived high threat level, a focus on force protection, and specific resources allocated to civil–military cooperation, such as on-site Taekwondo training for locals, ultimately aimed at improving the image of South Korean peacekeepers.[18] Finally, international expectations for Korea to contribute to international peace and security certainly play a role as well,[19] especially when voiced by a Korean at the helm of the UN, the former secretary-general Ban Ki-moon, who is also a former South Korean minister of foreign affairs and trade. Lee and Park nod further to a study that links the support for PKO deployments to the ambition to support one of their own as secretary-general.[20]

A further political motivation for an active engagement with global governance, in general, and peacekeeping, in particular, is seen in the rivalry for international support with other East Asian states, most prominently China and Japan.[21] For South Korea, the People's Republic of China (PRC), as a permanent member and thus veto power in the UN Security Council, and Japan, as a major contributor to the UN budget, are both already well-established rivals within the UN framework in a race for international support.[22] Beijing's position on peacekeeping, however, remains rather ambiguous because, despite a role as staunch defender of the noninterference principle, it is still deploying peacekeepers to resource-rich areas in Africa. Japan, on the other hand, with a pacifist constitution and pacifist constituencies, remains largely uneasy about overseas troop deployments.[23] Against this background in Korea, a proactive peacekeeping profile is seen as an opportunity to demonstrate a committed middle-power role, with China and Japan being restricted to the sidelines. International observers now attribute to Korea a growing status and influence as a "rising middle power," with the country "becoming more of a provider of global security than solely a consumer."[24] Under the Global Korea national security strategy of President Lee Myung-bak, especially, peacekeeping played a central part in Seoul's aspiration for "middle power activism"[25] or in the ambition to "embrace international responsibilities and actively contribute to resolve global challenges."[26]

The following sections are now not so much an assessment of the performance or efficiency of South Korean peacekeeping but instead are focused on how major peacekeeping deployments during the presidencies of Roh Moo-hyun, Lee Myung-bak, and Park Geun-hye have been legitimated in the wider domestic political discourse. These are, basically starting in 2006, the contributions to UNIFIL and UNMISS, as well as the "United Nations Peacekeeping Operations Participation Act." Because the wider political discourse consists of the government, in this case the president and the respective ministries of foreign affairs and national defense, as well as the political opposition, this approach foregrounds domestic contestation (especially in the National Assembly), as well as changes over time. At the center of the analysis are questions about which specific role conceptions and self-images have been used by different political actors to justify the peacekeeping deployments and legislation.

Role Conceptions, Key Themes, and Significant Others

There are a number of retrieved role conceptions that have been employed in the South Korean peacekeeping discourse from 2006 to 2016 throughout the political spectrum: the dominant one being "responsible member of the international community," with minor ones like "guardian of world peace" and "global model nation," as well as the main rival role conception of "anti-militarist power." The main role conception over the whole observation period was the little contested and thus hegemonic "responsible member of the international community," championed by the Ministry of Foreign Affairs and Trade, which was complemented by the minor "guardian of world peace" role conception put forward by representatives from the Ministry of National Defense and the minor "global model nation" role conception pushed for by the Lee Myung-bak administration. The sole rival role conception is titled "anti-militarist power," which has been put forward not only by minor parties from the political left, but also from progressive lawmakers from the main opposition party, in its various platforms as Uri Party, Minjoo, or Democratic Party (DP). The three role conceptions will be dealt with separately in the following sections.

Overall, there was broad bipartisan support for the participation in UN peacekeeping missions since 2006, as illustrated by the two following tables. Table 5.1 is a list of the votes against the "United Nations Peacekeeping Operations Participation Act,"[27] which was tabled in the National Assembly on 29 December 2009 and was approved by 64.8 percent of lawmakers. The law stipulates a clear and transparent institutional process for contributing South Korean troops only to UN missions and ensures the parliament's rights for the approval, extension, and cancellation of deployments.[28] It also allows for the establishment of standing forces, an aspect that was met with significant criticism during the debate in the National Assembly. Out of 199 votes, 129 were in favor of the act, with 54 "no" votes and 16 abstentions.[29] This can be considered broad support for Seoul's PKOs profile. The "no" votes, however, came from all parties then present in the parliament, from the ruling conservative Grand National Party (Hannaradang; GNP) over the main opposition party, the progressive DP (Minjoodang), to the far-left Democratic Labor Party (Minju Nodongdang; DLP). Also noteworthy is that half of the representatives from the progressive DP did not support the PKOs bill. After all, the DP is the main opposition party and has had its candidates successfully run for president, as in the cases of Kim Dae-jung and Roh Moo-hyun.

When looking into the plenary debate about the PKOs bill in the National Assembly, the main points of disapproval concerned the perceived attempt

Table 5.1 "No" Votes and Abstentions on the "United Nations Peacekeeping Participation Act," According to Party Membership

Overall 199 votes
YES 129 (64.8 percent)

NO 54	**Abstentions 16**
Conservative:	Conservative:
Grand National Party 6/153	*Grand National Party* 5/153
Liberty Forward Party 7/18	*Future Hope Alliance* 3/14
Future Hope Alliance 1/14	Progressive:
Progressive:	*Democratic Party* 8/81
Democratic Party 31/81	
Democratic Labor Party 4/5	
New Progressive Party 1/1	
Creative Korea Party 1/3	
Independents 3/25	

Source: National Assembly Minutes of the 18th National Assembly, 285th Meeting and 1st Session, Plenary Session. National Assembly, 29 December 2009.

by the government to limit parliament's constitutional right in the approval of overseas deployment potential, as well as the risk of being drawn into international crises once standing units had been set up. Representative Cho Seung-soo from the minor left-wing NPP, for example, warned his colleagues that quicker approval guidelines might be abused by the government to pressure the National Assembly with reference to international expectations into agreeing with their deployment proposals.[30] The conservative lawmaker Park Seon-yeong from the minor Liberty Forward Party compared voting for the bill to "committing suicide as parliamentarians."[31] Another explanation for the high number of "no" votes or abstentions might be found in the fact that the issue of overseas troop deployments was politically still sensitive after the more controversial decision to contribute the Zaytun Unit to the US-led "coalition-of-the-willing" MNF in Iraq from 2004 to 2008. The support for the Zaytun Unit's deployment to Iraq was orchestrated by the progressive President Roh Moo-hyun, despite criticism from his own party, as a necessity in order to strengthen the US–ROK alliance. This initial hesitance and disapproval by parliamentarians concerning PKO deployments did fade, however, as can be seen in Table 5.2. The approval for the extensions of UNIFIL and UNMISS deployments was consistently high, despite left-wing parliamentarians' disapproval and for some years even without any debate before the voting in the National Assembly.

The Hegemonic Role Conception: "Responsible Member of the International Community"

Throughout the whole observation period from 2006 to 2016, the South Korean debate about peacekeeping deployments was governed by a hegemonic role conception, generally supported by conservatives and progressives, but ultimately not by the political far-left. This role conception can be

Table 5.2 Parliamentary Approval of UNIFIL and UNMISS PKO Deployments, 2006–2015

Date (NA-Meeting-Session)	*Bill*	*Overall Votes*	*Yes*	*No*	*Abstention*	*Yes in %*	*No in %*	*Abstention in %*
2006.12.22 (17–263–1)	UNIFIL Deployment Bill	175	119	34	22	64.3	21.1	12.8
2008.07.06 (18–276–5)	UNIFIL Extension	199	187	10	2	94.0	5.0	1.0
2009.07.15 (18–283–1)	UNIFIL Extension	245	221	11	13	90.2	4.5	5.3
2011.12.29 (18–304–2)	UNIFIL Extension	149	145	3	1	97.3	2.1	0.7
2012.09.27 (19–311–9)	UNMISS Deployment Bill	159	113	27	19	71.1	17.0	11.9
2012.11.22 (19–311–12)	UNIFIL Extension	154	118	14	22	77.1	9.8	13.1
2013.12.19	UNMISS Extension	199	179	13	7	89.9	6.5	3.5
(19–321–1)	UNIFIL Extension	194	171	13	10	88.1	18.3	5.2
2014.12.02	UNMISS Extension	249	214	17	18	85.9	6.8	7.2
(19–329–13)	UNIFIL Extension	251	218	14	19	86.9	5.6	7.6
2015.11.30	UNMISS Extension	202	183	6	13	90.6	3.0	6.4
(19–337–12)	UNIFIL Extension	210	192	6	12	91.4	2.9	5.7

Source: Own compilation based on National Assembly Minutes.

labeled "responsible member of the international community" after the wording (*gukjesahoeui irwon*) regularly used by the different administrations. In this discursive formation, duty, and responsibility that come with the status of an advanced country on the global stage are central aspects. Starting with the first debate about the deployment of the Dongmyeong Unit to UNIFIL in the National Assembly Standing Committee on Unification, Foreign Affairs and Trade on 5 December 2006, this trope was at the core of the legitimation narratives for peacekeeping deployments by all three administrations, whether conservative or progressive.

For example, in that first meeting on UNIFIL, then first vice foreign minister Jo Jung-pyo justified on behalf of the government the deployment of troops to Lebanon as a "responsibility in the international community"[32] and an "opportunity to show strong commitment to Lebanon and to world peace."[33] This was later reiterated in the plenary debate on 22 December by Representative Im Jong-seok from the ruling Uri Party, the main progressive party that later morphed into various DP outlets, when he stated that UN peacekeeping corresponds with Korea's international position and is thus in the national interest.[34] There are more examples: two-and-a-half years later, in July 2008, conservative lawmaker Representative Hwang Jin-ha, nicknamed the "Peacekeeping Preacher"[35] because of his strong advocacy for South Korean peacekeeping, of the now-ruling GNP, finished his special committee's report on the extension of the UNIFIL deployment with the notion that "our country is determined to be a responsible UN member."[36] Moreover, in July 2009 Foreign Minister Yu Myung-hwan argued for another extension of the UNIFIL mandate "as a member of the UN"[37] and in 2013 his successor as Foreign Minister Yun elaborated that as a nonpermanent member of the UN Security Council, a position Seoul held from 2013 to 2014, ongoing support for the PKO deployment is expected of Korea and is thus appropriate.[38] Last, the Ministry of National Defense refers to Korea as a "responsible member" in its Defense White Papers in 2010, 2012, and 2014 with regard to the country's participation in UN peacekeeping.[39]

Furthermore, this role conception was exhibited by Presidents Roh Moo-hyun and Lee Myung-bak, as will be shown in the following selection of statements. In order to assess the larger meaning of these references, it is insightful to have a closer look at the framing of the "responsible member" trope. In his State of the Nation address presented to the National Assembly on 6 November 2006, for example, President Roh announced that "in accordance with its elevated international status, Korea will faithfully carry out its role as a responsible member of the international community."[40] And in his New Year's address to the nation in 2009, President Lee Myung-bak remarked:

> The Republic of Korea is not the marginal state it was in the latter part of the 19th century when it suffered from the whims of international currents; nor is it the flickering candle that it was during the foreign exchange crisis. The country is emerging as a responsible member of the leading nations helping create a new international order.[41]

These two quotes by Korean presidents stand as representative of two main points about the "responsible member" narrative.

First, highlighting South Korea's responsibilities in the international community is closely linked to the acknowledgment of Seoul's risen international status. The prevalence of the argument in the South Korean domestic debate about peacekeeping deployments speaks for its effectiveness as a legitimation strategy, as it obviously resonates with the participants of the foreign policy discourse. Anyone who agrees with the image of Korea holding an "elevated status" and being "not marginal" anymore remains susceptible for expectations to play the appropriate international role as a PKO-deploying nation. Second, against the background of a modern history of marginalization and domination by foreign powers, the prospect of being a member of "leading nations" by exhibiting the responsible or appropriate behavior for one's international status raises the likelihood for external recognition as "one of them." Participating in the peacekeeping practice can be considered a safe route to external recognition as a committed country in the thick of global affairs. In the Lee Myung-bak quote from 2009, however, there is also an indication of a growing ambition to lead together with others and shape the future order. As will be shown shortly, there is more to be said about this leadership ambition.

Drawing From History

There are two further aspects of the "responsible member of the international community" role conception that are worth noting: the selective use of historical references and the keen ambition to foster and not damage diplomatic standing and influence. The duty or responsibility to participate in UK PKOs is not only linked discursively to the reassurance of status but also very concretely to the narrative that South Korea is a "child of the UN" because of UN support during and after the Korean War. Government officials and representatives of the ruling parties, especially, refer to this "historical debt" toward the international community (and the UN, in particular) that can be paid back through peacekeeping efforts. These historical references are a constant feature of the "responsible member" role conception, regardless of party ideology or time period. For example, in December 2007 Foreign Minister Song Min-sun emphasized the importance of this "payback" in a television interview with the Korean Arirang Channel:

> We were helped by the international community half a century ago. If we were not helped by the international community, nobody could have enjoyed this peace and prosperity in which we are now living. It is time for us to requite, to return the benefits we got from the international community to the world.[42]

Historical debt was also noted by Representative Jeong Ok-im from the conservative GNP as a reason why Seoul should continue to deploy peacekeepers to UNIFIL during the plenary debate on 15 July 2009.[43] Another example is the 2010 statement of Vice Foreign Minister Shin Kak-soo at a luncheon at the Korean UN Association in Seoul on UN Day, 25 October: "As a country that owes much to the UN, Korea has embarked on a path to contributing actively to the world community."[44] There is, however, more to this historical narrative, as it is not simply focused on historical debt but also on historical achievement. Take, for example, President Lee's Address to the Nation in November 2009, where he adds the following passage after acknowledging the historical support from the international community:

> The Republic demonstrated that the assistance and collaboration from the international community was never wasted and that it was only right for such aid to be further expanded. In this regard, Korea was an exemplary beneficiary of international aid and cooperation in the 20th century. In the 21st century, the nation stands as a beacon for developing countries. Now is high time for the nation to return in full what it owes to the international community. Among other things, the Republic intends to share what it has experienced in achieving success with other countries across the globe.[45]

Here, the experience of the Korean War and subsequent reconstruction becomes a "unique experience"[46] that is supposed to serve as a model for other conflict-ridden countries while South Korea is capable and willing to help and assist. Thus, South Korea is not only acting out of a perceived duty to pay back a historical debt but also because Seoul sees itself as uniquely positioned to be a role model in the fields of development assistance and peacekeeping. Foreign Minister Yu Myung-hwan, for example, expressed this explicitly in 2008: "We believe that our own experiences of rebuilding the nation in the aftermath of a devastating war enable us to play a unique role in these areas."[47] In 2015 the director for international policy at the Ministry of National Defense described this as a development "from beneficiary to benefactor."[48]

Whereas peacekeeping deployments have been rendered meaningful and appropriate as a grateful duty in order to repay historical debt, the focus on national achievements in terms of development and defense makes peacekeeping a proud display of economic and military might as well as of a certain generosity. According to this framing, Korea is not so much historically obliged to engage in UN peacekeeping but is uniquely qualified to share its own experience and viewpoints with the world because Korea has something to offer to the world itself. President Park Geun-hye also tried to evoke this spirit of national achievement, for example, in a statement on the 2015 New Year:

> Having been one of the poorest countries in the world after the Korean War, Korea has risen to become [one of] the world's major economies and the only country to have successfully accomplished heavy chemical industrialization without going to war. Ours is the first country in the world to have gone from aid recipient to aid donor. Such is the strength of the Korean people, which is why I am confident we will be able to overcome whatever difficulties lie before us.[49]

The perspective is not just present under conservative leaders. During the progressive Presidency of Roh Moo-hyun, Foreign Minister Song Min-sun in 2007 as well as Vice Foreign Minister Jo Jung-pyo in 2006 already made the case for Korea's unique potential to contribute to international peace and security, especially in the Middle East as a non-Muslim country that shares the experience of being colonized, in contrast to many other major powers.[50]

Among the three analyzed administrations, the Lee Myung-bak administration stands out in its frequent use of the "beneficiary to benefactor"[51] historical narrative. Against the background of President Lee's ambitious Global Korea national security strategy this is hardly a surprise. This national project was set up to transform Korea into an "advanced and globally prestigious country"[52] in the eyes not just of the world but also Koreans themselves. In one of his last major speeches before the end of his presidency in February 2013, Lee Myung-bak reiterated what he had told Koreans often before:

> In 2011, a renowned foreign magazine evaluated Korea, noting that long ago the world had taken note of Korea as a model for growth and newly

> emerging power, but that Koreans themselves did not seem to be aware of the fact.[53]

Although President Roh had also repeatedly encouraged Koreans to have more confidence in the country's potential and achievements,[54] President Lee's vision went further in suggesting a clear path for how to achieve a Korea that is at the center and on top of global affairs, a Korea that is leading rather than being led, and a Korea that is inspiring other nations to follow suit. In his first National Address on Liberation Day, 15 August 2008, he told the country about a dream that he framed as the dream of all Koreans:

> We are all dreaming of an advanced nation standing tall in the international community where the happiness of each individual goes hand in hand with the progress of the nation and where material abundance strikes a balance with mental maturity. It will be a strong nation of an affluent people and a compassionate society. Such a nation will serve as a paragon for humankind and will be respected by other countries in the international community.[55]

Clearly, President Lee formulates a goal here for a future Korea as an "advanced nation" (*seonjinguk*), which he suggests can be realized through the Global Korea strategy. This strategy stands for global outreach and more international engagement, with peacekeeping as one of the key policy areas. However, there are two slightly different goals formulated in the quoted paragraph, one about the aspiration to maturity and the other about the ambition to lead.

First, there is the aspiration to become a mature global country (*seongsukan segyegukga*). Mature behavior here may be understood as a foreign policy that is not simply about acting "selfishly" but as an attitude beyond self-interest that contributes to the "common good." In a speech on nation branding from 2013, President Lee makes clear, however, that a mature foreign policy behavior is not a value in itself but serves a further purpose:

> The principle that works in the improvement of an individual's reputation can also be applied to a nation's brand value. People do not show respect to those who are simply rich, but rather to those who share what they have and serve others. In addition, people have a tendency to be fascinated by those who exhibit sophisticated cultural maturity. The same is true for a nation. A country can win global confidence only by becoming a mature member of the international community that shares and gives help to other nations going beyond establishing itself as an economic powerhouse.[56]

This rare statement reduces maturity explicitly to a way of fashioning a certain brand and of gaining trust on the global stage, because the respective practices are seen as conducive to gaining a higher status. In the literature on norm socialization, such an instrumental perspective would be seen as "Type

I socialization" achieved through what Checkel called—in a, for role theorists, simplistic understanding—"conscious role playing": agents are simply "knowing what is socially accepted in a given setting or community" without having norms internalized completely and without seeing them as "the right thing to do."[57] From most other less explicit statements in the peacekeeping discourse, however, one cannot draw similar inferences about the exact understanding of maturity held by various other political actors. In the end, the question of whether President Lee has internalized international norms and actually believes in contributing to the common good is less relevant for the analysis at hand. What matters here is how specific actions have been framed as appropriate, meaningful, and legitimate with regard to collective identity narratives, as these interactive structures of meaning underpin and constitute foreign politics as a whole.

Second, President Lee formulated the clear ambition not just to commit to the global order but also to shape international affairs through displaying leadership. This is very clear in his address on 15 August 2012, Liberation Day, when the president was reflecting on the global power shift from West to East after the Global Financial Crisis (GFC):

> At this important juncture, Korea has just entered the ranks of advanced nations breaking away from the group of late starters that it had long belonged to. From now on, it is not viable to follow others. Our country is now in a position to open a path on its own. We have to develop "a Korean route" that no one else travelled before. [. . .] The strength that we have had so far as a runner-up may ironically become a stumbling block as a frontrunner.[58]

Note the parallel between the "from runner-up to frontrunner" and the "from beneficiary to benefactor" narratives. President Lee's envisioned Global Korea is not simply content with committing itself to the status quo international order but has to lead the world itself. This leadership ambition based on historical achievements could be seen as the core of yet another role conception, that of a "global model nation." In the peacekeeping context, this role conception is only expressed by the representatives from the Lee Myung-bak administration and is complementing rather than contesting the main role conception of "responsible member of the international community." Still, in 2013, Representative Hwang Jin-ha from the conservative New Frontier Party (Saenuridang; formerly the GNP), a retired ROK Army Lieutenant General who himself was a force commander of the United Nations Peacekeeping Force in Cyprus (UNFICYP)[59] from 2002 to 2003, and an influential lawmaker on the issue, confronted Foreign Minister Yun Byung-se over what Hwang called lack of ambition over finally making a Korean commander of UNIFIL, after deploying units to Lebanon for over six years, to the positive appraisal of the UN as well as the Lebanese government. Furthermore, Hwang said he was "embarrassed."[60] Foreign Minister Yun agreed that there should be more Koreans serving at the UN in high-ranking positions.[61] In 2015, this aspect was reaffirmed by Colonel Ha Jong-sik from the Overseas Deployment Division at the ROK Joint Chiefs of Staff: "South Korea will dedicate significant diplomatic efforts to ensure that its officials have senior-level PKO positions [. . .] corresponding to South Korea's contribution to UN PKO activities."[62] Another minor role conception found is that of "guardian of world peace," mostly used by officials from the Ministry of National Defense and more intended for the global role of the ROK military than for the whole country. Take for example this other statement made by the same official from the ROK Joint Chiefs of Staff:

> During the Korean War, Korea received aid from many countries around the world, but now it has become a donor that provides support to other countries. As such, the 634 troops currently carrying out their duties as part of seven PKO missions will fulfil their role as guardians of world peace.[63]

This role conception, too, does not stand in conflict with the main role conception of "responsible member of the international community" but rather complements it in its commitment to the status quo international order. What is common to all three role conceptions is that the significant other is the rather abstract international community, often simply represented by the UN, which may not be that surprising at all as this analysis is concerned with UN peacekeeping. Still, it is the perceived appropriate behavior and expectations from this international sphere that are used by political actors to justify their commitment to the peacekeeping deployments.

Raising International Standing and External Recognition

Besides historical references, international standing is the other core aspect of the "responsible international member" role conception in the South Korean political discourse on peacekeeping. It remained a constant feature of the government's case for peacekeeping deployments, which was also framed as an opportunity to further raise Seoul's international status, standing, or brand. For example, on the website of the ROK Permanent Mission to the UN it reads: "The level of Korea's participation in PKO is a reflection of the government's willingness to contribute to world peace and security, thus enhancing its status in the international community while simultaneously making the world a safer place."[64] In the 2012 Defense White Paper it says:

> Through the expansion of global peacekeeping activities, the ROK Armed Forces are fulfilling our responsibility to the international community on a level that is befitting our national power. In the process, we are enhancing our global stature as a defender of world peace.[65]

During the 2009 debate over the extension of the UNIFIL extension in the National Assembly Standing Committee on Foreign Affairs, Trade, and

Unification, Vice Minister of Foreign Affairs Shin Kak-soo promised that "the government will continue to make efforts to enhance the international status of Korea by actively contributing to the UN's efforts to maintain international peace and security."[66] In a similar but ultimately more insightful way, parliamentary senior expert Seong Seok-ho explains the government's proposal to extend the mandates of the Korean deployments to UNIFIL and MINUSTAH:

> Through the extensions, we hope to contribute not only to peace and stability in the region, but also to secure a strong base for diplomatic ties between those countries and Korea as well as to expand our influence in the international community through military contributions corresponding to our financial support.[67]

From this exemplary statement it becomes clear what is at the core of nearly all justifications of peacekeeping deployments coming from the Lee Myung-bak administration: the belief that the PKO deployments will help to foster diplomatic ties and thus increase Korea's influence around the world. Again, this is very much in line with the global outreach envisioned by the Global Korea strategy, as the following statement by Foreign Minister Kim Sung-Hwan illustrates: "[N]etworks are emerging as a key element in the diplomacy of the 21st century. Achieving the goal of a global Korea will also depend on how effectively we can build networks of cooperation with countries around the world."[68]

This diplomatic campaign is supposed to be achieved through the successful establishment of an international perception of Korea as "a responsible member of the international community." However, the promises and pitfalls of this military diplomacy became visible in quite practical terms in two events in 2008 and 2013. On the one hand, during the initial review of the UNIFIL deployment, the Joint Chiefs of Staff operations division chief Jang Gwang-il reported how Korea's reputation was not only successfully improved among Lebanese locals through Taekwondo and Korean-language classes but also through the establishment of personal ties with other peacekeeping units deployed there. When in 2008 "one of the largest earthquakes in human history in terms of socio-economic losses"[69] devastated the Chinese province of Sichuan, the Korean peacekeepers collected donations and delivered them to a neighboring Chinese peacekeeping battalion.[70] On the other hand, in late 2013 news broke about a controversy between Korea and Japan over an incident in South Sudan, where, in a tense security environment, South Korean peacekeepers from the Hanbit Unit had requested additional ammunition from the UNMISS command and were handed bullets from Japanese peacekeepers. Back home, the Japanese opposition claimed that arms-export rules had been violated, and the Korean side refuted reports that the Hanbit Unit had made a direct request to its Japanese peers, attributing the whole story to a malicious spin by Prime Minister Abe's administration.[71]

The importance of diplomatic ties and how to avoid damaging them with peacekeeping deployments had already been an issue under President Roh

Moo-hyun in 2006, when the UNIFIL deployment was first debated in parliament. Whereas Vice Foreign Minister Jo Jung-pyo argued that the Lebanon engagement would ultimately help foster diplomatic ties with Arab nations, because the Korean reputation in the region had suffered after supporting the US in Iraq with the Zaytun Unit,[72] Representative Im Jong-in from the ruling progressive Uri Party feared that the damaged "brand" of Korea would not be improved by such a measure and only endanger Korean soldiers there.[73] As will be shown in the next section about the main rival role conception, this purported diplomatic benefit from peacekeeping deployments was contested not only by parties from the political left but also by the selective use of historical references as well as the perception of the international community.

The final aspect of the "responsible member in the international community" role conception is closely related to the goal of raising Korea's international standing: external recognition. In what can fittingly be illustrated with Putnam's logic of the two-level game,[74] South Korean decision-makers must negotiate their roles not only domestically, which is the focus of the analysis at hand, but also internationally with other international actors. Accordingly, through their commitments to international peacekeeping efforts, South Korean decision-makers aim to become recognized as "responsible members of the international community" while establishing networks for further projection of national influence, as shown earlier. On the other hand, however, international recognition of a certain status and the related expectations for appropriate behavior can serve as a welcome legitimation strategy to garner domestic support. If South Korean decision-makers and citizens want to continue seeing their country holding a certain status, they have to support the prescribed behavior or role performance. The fulfillment of these expectations, in turn, validates the role-playing country's status aspirations and ambitions. It is no wonder then that a key feature of any parliamentary debate about peacekeeping mandate extensions is based on the government's report about the various praise the deployed units and the whole country received locally as well as internationally.

For example, in the 2008 Defense White Paper, the Dongmyeong Unit reportedly was praised as "Godsend"[75] or "a family, a brother"[76] by the Lebanese locals. Furthermore, the unit was awarded a UN Medal and hailed by the UNIFIL force commander as an "exemplary unit involved in peacekeeping operations."[77] President Lee Myung-bak told Koreans on the patriotic March First Independence Movement Day in 2012 that "[o]ur Armed Forces have won the hearts and minds of local residents more than any other country's military forces. They are thus receiving high appraisal internationally."[78] In 2014 a vice foreign minister made the case for a further UNIFIL mandate extension as follows:

> The United Nations and the Lebanese government highly appreciate the accomplishments of our united forces and are requesting continued dispatch, and it is hoped that the extension of our dispatch will enhance our

> international status and role and contribute to the development of bilateral relations between Korea and Lebanon.[79]

According to Lee Jeongkyu, director-general for the International Policy Bureau in the Ministry of National Defense, the UNMISS force commander also praised one operation by the Hanbit Unit as "one of the most successful operations carried out for the protection of refugees in the history of the UN Peacekeeping Forces."[80] Lee further reports that

> [a]t first, locals took an aloof attitude toward the presence of the ROK troops, but they have gradually opened up in response to the sincere humanitarian aid and services that the troops have provided. Now, they shout "Korea wonderful!" and give a thumbs-up to members of the Hanbit Unit when they pass by. Some of them even greet the troops with basic Korean words.[81]

These reported successes in military diplomacy are not what critics of South Korean peacekeeping deployments had feared a decade ago. Back then, criticism was coming from the political left, which contested not only the selective use of history to legitimate peacekeeping deployments but also, more generally, troop deployments and military diplomacy as appropriate means for global engagement, as will be shown in the next section.

The Rival Role Conception: "Antimilitarist Power"

The final and main rival role conception could be called "antimilitarist power." This role conception was exhibited by a number of left-wing opposition parties, such as the DLP and the splinter NPP, which both merged again later as the Unified Progressive Party (UPP). But related concerns about peacekeeping deployments were also voiced by individual lawmakers from the main progressive opposition party, the Minjoo, or DP, which was formerly the Uri Party under the progressive presidency of Roh Moo-hyun. This role conception remained marginal but present throughout the observed period and may hold some explanatory value also for the abstentions regarding peacekeeping deployments by significant numbers of more mainstream DP lawmakers.

A key theme for this rival role conception is also the use of historical narratives, but whereas the Korean War and the subsequent Korean development were central to the "responsible international member" role conception, for the political left the relevant historical events are the contentious deployment of troops to two wars fought alongside the US, Vietnam, and Iraq. This is very much in line with what has been termed the tension between the "Vietnam syndrome"[82] and the "child of the United Nations" narrative in the literature on South Korean peacekeeping. Whereas the former is a discourse about futile military adventures for the sake of the US–ROK alliance, the latter emphasizes the historical debt of South Koreans toward international support during and after the Korean War.

The left-wing lawmaker Cho Seung-soo from the NPP, which separated itself from the DLP in 2008, illustrates the "Vietnam syndrome" during the plenary debate on the mandate extension for deployed South Korean peacekeepers to Lebanon in July 2009 and in the debate about the UN Peacekeeping Participation Act in December 2009. In addition to introducing Korea's involvement in the Vietnam War as a negative lesson from history ("We already have such an experience in history when we sent tens of thousands of soldiers to Vietnam and inflicted pain on the Vietnamese people. This was due to our US-centered foreign policy"),[83] Cho directly contested the argument that "historical debt" translates into a responsibility to contribute military troops to international peacekeeping efforts. Instead, he used the collective experience of the Korean War to paint the image of Koreans as "our peace loving people" (*pyeonghwareul saranghaneun uri gungmindeuri*) who are shamefully being led into war by the main two parties, the conservative GNP and the progressive DP.[84] To him the envisioned fast deployment capabilities through the establishment of standing units were not sending the right image of Korea as a "peaceful country" (*pyeonghwagukga*) to the Middle East, risking the perception that South Korean soldiers will again be seen as "mercenaries of the United States" (*migugui yongbyeong*).[85] He stated that this misperception ultimately leads to less safety for Korean soldiers and citizens around the world. Furthermore, for Cho, the fast deployment capabilities from the PKO Act not only violated the rights of the National Assembly to examine and approve overseas troop deployments but also the peaceful spirit of Article 5 of the Korean Constitution, which prohibits wars of aggression.[86] But lawmaker Cho also reaffirmed that his party was committed to international development, and saw the majority of ordinary Koreans behind his position.[87] Finally, Cho linked the historical Korean participation in the Vietnam War with today's UN PKOs in order to frame present-day peacekeeping deployments as a distortion of Korea's peaceful nature.

In a similar way, in 2006 in the debate about the first UNIFIL deployment, Representative Im Jong-in from the then-ruling Uri Party had warned that the Korean image in the Middle East, which had already suffered from the Korean involvement in the Iraq War, would only deteriorate further, damaging the Korean brand (*hanguk beuraendeu*) even more.[88] For him the decision about the deployment proposal was clear: "Please stop this motion and raise our country's prestige again."[89] This is in stark contrast to the "responsible member" role conception, where discursive agents construed an improved image and heightened prestige in the "international community" as a result of peacekeeping commitments with boots on the ground. Common to both conceptions, however, is the concern for Korea's international standing and diplomatic ties, especially to countries in the Middle East. Another example of this aspect is DLP lawmaker Gwak Cheong-suk's concern that relations with Arab nations might suffer over the deployment.[90] In addition, he accused the government of hiding costs for PKOs from the National Assembly and doubted that the peacekeeping had any effect whatsoever on the security situation in Lebanon.[91] While he agreed that Korea has the mission to work toward international peace, he saw this goal only as served through a withdrawal and not further deployment of forces.[92]

Likewise, in 2012 Representative Kim Mi-hee from the UPP, a merger of various left-wing parties in succession of the DLP and prohibited by the Constitutional Court in 2014 for pro–North Korean views, called the peacekeeping mandate of the Hanbit Unit into question. For her, having combat troops in South Sudan to fulfill mainly reconstruction works and medical services was not conducive to the local security situation, as it might lead to an escalation of the local conflict: "The dispatch of Korean troops can intensify the dispute in South Sudan and not contribute to peace."[93] Instead of military troop deployments, she called for "unconditional and humanitarian support"[94] while suggesting that most other international powers, from the US and China to Japan, hold a presence in South Sudan only because of its rich natural resources.

This is illustrative of a certain skepticism toward the "international community" among left-wing politicians. For Cho Seung-soo from the NPP, international politics is usually governed by great power logics, which led him to call for Korea to balance for peace during the debate on the PKO law in 2009.[95] DLP lawmaker Hyeon Ae-ja described the UN Security Council Resolution 1701, which calls for the disarmament of Hezbollah, the Shiite political party and militant group whose war with Israel in July 2006 led to the UNIFIL II deployment in the South of Lebanon, as one-sided in favor of Israeli and US interests.[96] Furthermore, she criticized the "international community" for staying silent "when Lebanon was helpless against Israel's bombing."[97] The significant Other in the rival role conception "antimilitarist power" is not the international community but the US, given Korea's history of military deployments in the context of the US–ROK military alliance as well as the US interventionist foreign policy in the Middle East. It is against this background that the political left in Korea tries to reestablish a self-image of Korea as a peaceful country that is not taking part in resource imperialism, as the DLP's Kim Mi-hee was suggesting.[98] This leads the analysis at hand to the final section about the self-images at play in the discursive contestation over the different role conceptions with regard to South Korean peacekeeping.

Self-Images

After identifying the main role conceptions, key themes, and significant others in the South Korean wider foreign policy discourse on peacekeeping deployments, the final step of the analysis consists of a close, intertextual, and interpretative reading of the utilized self-images, as well as their commonalities and differences. Through these pixelated snapshots of identification practices, one can in a next step acquire a more comprehensive understanding of the underlying identity narratives informing and being informed by the global foreign policy actor, South Korea.

The first self-image prevalent in the peacekeeping discourse is closely linked to the hegemonic role conception "responsible international member" as well as the minor role conception "guardian of world peace." It is the image of a risen South Korea, a once peripheral and marginal country that has now entered the ranks of advanced, relevant countries and stands tall, proud of its achievements, on the global stage. Korea here seeks the global spotlight and is eager to contribute to the international community in order to acquire self-assurance through external recognition as "one of them," a privilege that has been withheld from Koreans since the turbulent end of the Joseon dynasty amid the dawn of "the international" in East Asia. This is well illustrated by the statement quoted earlier from President Lee Myung-bak's 2009 New Year address:

> The Republic of Korea is not the marginal state it was in the latter part of the 19th century when it suffered the whims of international currents. [. . .] The country is emerging as a responsible member of the leading nations helping create a new international order.[99]

The global outreach is thus a way of raising the visibility of Korea's journey from the periphery to the center, one might say back to a position close to the center that Joseon Korea once occupied in the Sino-centric world order before Western encroachment. The "beneficiary to benefactor" historical narrative that is central to the "risen Korea" self-image is a testament to this perspective. Moreover, the significant Other here is not North Korea or the US but the past Korea, the marginal state or the beneficiary. All in all, PKOs are a suitable way of forging ties with various international partners in a policy and diplomatic area of comparatively high international legitimacy, bestowed through the UN system and the obligatory invitation for UN PKOs from the host country and despite the inherent risk for the safety of deployed military troops, the small costs that are facilitated through the compensations from the UN budget. It is noteworthy that there is a new element to this, one that is reflected by the emergence of the role conception of being a "global model nation." Here Korea has the unique potential and ambition to venture further, to shape global affairs from the center as a pioneer. President Lee Myung-bak's Liberation Day address from 2008 is a good example of this: "[An] advanced nation [. . .] will serve as a paragon for humankind and will be respected by other countries in the international community."[100] Proponents of the main rival role conception of "antimilitarist power" would not argue with President Lee's dream, which is an expression of the "leading Korea" self-image, but they would certainly differ on the ways and means through which Korea can achieve the aforementioned goal.

Their utilized self-image is that of Korea as a country of peace, which is an idea that had also been championed by President Roh Moo-hyun in 2007. In his address on the March First Independence Movement Day he laid down a vision of Korea that merged several of the earlier mentioned key themes:

> As a nation that has done no injury to another throughout history, Korea holds a sufficient moral ground and qualification to lead peace efforts in Northeast Asia. The geopolitical map also places Korea on an equilibrium point where it is able to maintain the peace structure. When Korea's strength reached a high tide, peace in [the] Northeast was preserved and when it slid down to a low ebb, the region was swirled into a maelstrom of war, a manifestation that the order of Northeast Asia depends on the action that this nation takes.[101]

This self-image, which can be called "righteous Korea," is inherently peaceful and a force for peace in the region. Moreover, Korea has the "moral ground" and potential to lead the region into an era of peace. This mirrors the remarks of NPP lawmaker Cho Seung-soo, who warned that the "peace loving Korean people"[102] might risk being misunderstood or misperceived as "mercenaries of the United States" by others over their growing peacekeeping profile in the Middle East, somehow against their true nature. Regarding Roh's historical claim of Korea's peaceful rise, it can be argued that this is at odds with Korea's involvement in the Vietnam War, which proved instrumental in funding South Korean economic development.

Finally, the left's image of a peaceful Korea is very much at odds with the mainstream self-image of a South Korea at the center of today's US-led international order. For the political left, the "peaceful Korean people" can and should stand aside when others are engaging in militaristic "resource imperialism" and neglecting the needs of ordinary, helpless people around the world. These left-wing advocates of a pacifist, humanitarian, or civilian South Korea are not calling for isolationism—and here is the common key theme to both major self-images and respective role conceptions—but for a widespread and largely nonconfrontational international outreach.

Conclusion

Overall, there has been broad political support for participation in UN peacekeeping missions since 2006. The main role conception over the whole observation period was the little contested and therefore hegemonic "responsible member of the international community," linked to a self-image of a risen South Korea, a once peripheral and marginal country that has entered the ranks of advanced countries and is eager to contribute to the international community in order to acquire self-assurance through external recognition. However, the duty or responsibility to participate in UN PKOs is not only linked discursively to the reassurance of status but also to the historical experience of being helped by the UN during and after the Korean War. In addition, South Korea is not only acting out of a perceived duty to pay back a historical debt but also because Seoul sees itself as uniquely positioned to be a role model in the fields of development assistance and peacekeeping. This leadership ambition, which is most at the core of President Lee Myung-bak's Global Korea strategy, is based on historical achievements that could be seen as the core of yet another role conception, that of a "global model nation"; this, in turn, is linked to the "beneficiary to benefactor"[103] historical narrative. The discursive agents behind the "responsible member" role conception also emphasized the image improvements and heightened prestige in the international community as a result of peacekeeping commitments with boots on the ground.

This concern for Korea's international standing and diplomatic ties especially to countries in the Middle East is common to both proponents of the dominant role conception as well as the major deviant role conception of "antimilitarist power." This rival role conception was put forward by representatives from the political left, who argued that historical debt does not automatically translate into a responsibility to contribute military troops to international peacekeeping efforts. Instead, they used the collective experience of the Korean War to paint the image of Korea as a peaceful nation opposed to war and to argue that Korea should stand aside when others, especially the US, are engaging in militaristic "resource imperialism." In consequence, the political left's image of a peaceful Korea is very much at odds with the mainstream self-image of "rising Korea" at the center of today's US-led international order.

Figure 5.1 is an attempt to visualize these self-identifications as links between role conceptions and self-images. Because of the stability of the self-identifications linked the self-images of "rising Korea" and "leading Korea" to the "responsible international member" role conception, one can conclude that there was indeed a dominant role identity throughout the observation period, that of "responsible international member." Through the focus on self-identifications, however, the analysis revealed how the role-playing was contested and which self-images were linked to which role conceptions in what way.

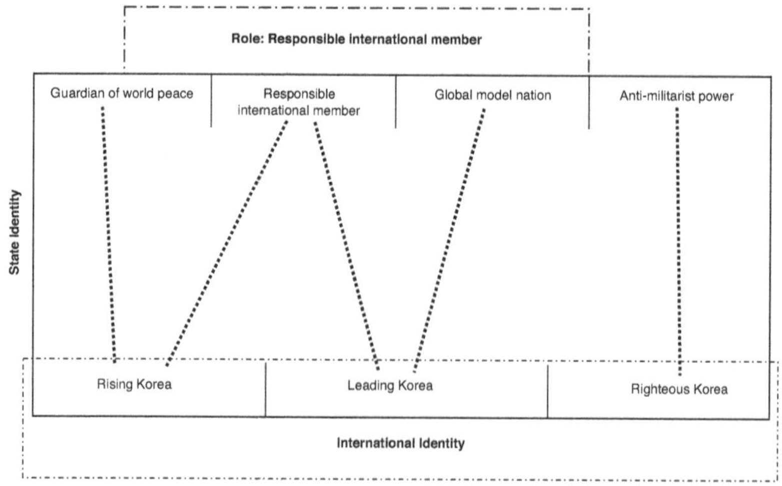

Figure 5.1 Roles, Role Conceptions, and Self-images in ROK Peacekeeping, 2006–2016

Source: Created by author.

Notes

1. Hwang, Balbina Y. 2012: "Korea and PKO: Is Korea Contributing to Global Peace?", in: Scott A. Snyder (ed.), *Global Korea: South Korea's Contributions to International Security*, New York/Washington, DC: Council on Foreign Relations, pp. 13–27; Ikenberry, G. John and Mo, Jongryn 2013: *The Rise of Korean Leadership, Emerging Powers and International Order*, New York: Palgrave Macmillan, p. 145; Heo, Uk and Roehrig, Terence 2014: *South Korea's Rise: Economic Development, Power, and Foreign Relations*, Cambridge: Cambridge University Press, p. 172.
2. Howe, Brendan and Kondoch, Boris 2014: "Northeast Asian Perspectives on UN Peacekeeping: China, Japan, Korea", *Journal of International Peacekeeping*, Vol. 18, No. 3–4, p. 146.
3. Hong, Kyudok 2009: "South Korean Experiences in Peacekeeping and Plan for the Future", *Advances in Military Sociology: Essays in Honor of Charles C. Moskos, Contributions to Conflict Management, Peace and Economics and Development*, Vol. 12A, p. 176.
4. Chung, Eun-sook 2010: "Korea's Law on UNPKO and Participation in International Peacekeeping Missions", *Korea Focus Current Issues and Policy, No. 166*. Available at: www.koreafocus.or.kr/design2/layout/content_print.asp?group_id=102934 [Accessed 5 November 2018]; Lee, Shin-wha and Park, Joon Sung 2014: "Contributor Profile: South Korea, International Peace Institute Peacekeeping Database", *providingforpeacekeeping.org*, 13 June. Available at: https://s3.amazonaws.com/ipi-pdf-document-store/ppp-profiles/asia/ipi-pub-ppp-Korea.pdf [Accessed 5 November 2018].
5. Lee, S.W. and Park, J.S. 2014.
6. Ministry of National Defense 2016: *Defense White Paper 2016*, p. 170.
7. Hwang, B.Y. 2012: p. 19.
8. Lee, S.W. and Park, J.S. 2014.
9. Ibid.
10. Sesay, Amadu 1997: "Dividends on Investments in History: Korea and UN Peace-Keeping and Peace-Building Operations", *Korea Observer*, Vol. 28, No. 2, pp. 181–217; Hong, K.D. 2009: p. 76; Lee, S.W. and Park, J.S. 2014; Howe, B. and Kondoch, B. 2014: p. 137.
11. Howe, B. and Kondoch, B. 2014: p. 137.
12. Hong, K.D. 2009: p. 173.
13. Cumings, Bruce 2005: *Korea's Place in the Sun: A Modern History*, New York: W. W. Norton Company, p. 321.
14. Baek, Glenn 2013: "A Perspective on Korea's Participation in the Vietnam War", *The Asan Institute for Policy Studies Issue Brief 53*, 9 April. Available at: http://en.asaninst.org/contents/issue-brief-no-53-a-perspective-on-koreas-participation-in-the-vietnam-war/ [Accessed 5 November 2018].
15. Breuker, Remco E. 2009: "Korea's Forgotten War: Appropriating and Subverting the Vietnam War in Korean Popular Imaginings", *Korean Histories*, Vol. 1, No. 1, pp. 40–41.
16. Hong, K.D. 2009: p. 182f.
17. Ibid.: p. 175.
18. Ruffa, Chiara 2014: "What Peacekeepers Think and Do: An Exploratory Study of French, Ghanaian, Italian, and South Korean Armies in the United Nations Interim Force in Lebanon", *Armed Forces & Society*, Vol. 40, No. 2, pp. 199–225.
19. Ikenberry, G.J. and Mo, J.R. 2013: p. 145.
20. Lee, S.W. and Park, J.S. 2014.
21. Hwang, B.Y. 2012: p. 20; Lee, S.W. and Park, J.S. 2014.
22. Howe, B. and Kondoch, B. 2014: p. 136.

23. Hwang, B.Y. 2012: p. 20.
24. Heo, U. and Roehrig, T. 2014: p. 166.
25. Ko, Sangtu 2012: "Korea's Middle Power Activism and Peacekeeping Operations", *Asia Europe Journal*, Vol. 10, No. 4, pp. 287–299.
26. Hwang, B.Y. 2012: p. 15.
27. A full text of the Act in an English translation can be found here: http://elaw.klri.re.kr/eng_service/lawView.do?lang=ENG&hseq=19640.
28. Lee, S.W. and Park, J.S. 2014; Chung, E.S. 2010.
29. National Assembly Minutes 2009: *Plenary Session 18-285-1*.
30. Ibid.: p. 17.
31. Ibid.: p. 16.
32. National Assembly Minutes 2006: *Standing Committee on Foreign Affairs Session 17-262-17*.
33. Ibid.
34. National Assembly Minutes 2006: *Plenary Session 17-263-1*.
35. Lee, S.W. and Park, J.S. 2014.
36. National Assembly Minutes 2008: *Plenary Session 18-276-5*.
37. National Assembly Minutes 2009: *Plenary Session 18-283-2*.
38. National Assembly Minutes 2013: *Plenary Session 19-320-4*.
39. See Ministry of National Defense 2010: *Defense White Paper 2010*, p. 106; *Defense White Paper 2012*, p. 3, *Defense White Paper 2014*, p. 143.
40. Roh, Moo-hyun 2006: *State of the Nation Address by President Roh Moo-hyun Presented to the 262nd Regular Session of the National Assembly*, 6 November.
41. Lee, Myung-bak 2009a: *New Year Policy Address to the Nation by President Lee Myung-bak*, 2 January.
42. Song, Min-sun 2007b: *Foreign Minister Song Min-soon's Interview with Arirang TV on 'Diplomacy Lounge'*, 29 December.
43. National Assembly Minutes, 2009: *Plenary Session 18-283-1*, p. 9.
44. Shin, Kak-soo 2010: *Keynote Address by H.E. Dr. Shin Kak-soo, Vice Minister of Foreign Affairs and Trade, at the Luncheon hosted by the UN Association of the Republic of Korea on UN Day*, 24 October.
45. Lee, Myung-bak 2009d: *Radio and Internet Address to the Nation by President Lee Myung-bak*, 30 November.
46. Kim, Bong-hyun 2013: *Statement by H.E. Ambassador Kim Bong-hyun, Deputy Minister for Multilateral and Global Affairs, the Ministry of Foreign Affairs and Trade of the Republic of Korea, Open Debate on 'Peacekeeping: A Multidimensional Approach'*, 21 January.
47. Yu, Myung-hwan 2008: *Speech by H.E. Yu Myung-hwan, Minister of Foreign Affairs and Trade, at the Luncheon Hosted by the UN Association of the Republic of Korea on the 63rd Anniversary of UN Day*, 20 October.
48. Ha, Jong-Sik 2015: "ROK's Achievements through Its PKO Activities and Future Prospects", *Korean Institute for Defense Analysis, ROK Angle 122*, 29 May.
49. Park, Geun-hye 2015: *Remarks by the President at the New Year Press Conference*, 12 January.
50. Song, M.S. 2007b; National Assembly Minutes 2006: *Standing Committee on Foreign Affairs Session 17-262-17*, p. 17.
51. Ha, J.S. 2015.
52. Yu, Myung-hwan 2009: "Diplomacy at Forefront to Realize Global Korea", *The Korea Times*, 1 March.
53. Lee, Myung-bak 2013: *The Nation Brand Value Soaring: From Korea Discount to Korea Premium*, 21 January.
54. Roh, Moo-hyun 2004b: *Address by President Roh Moo-hyun on the 59th Anniversary of National Liberation*, 15 August; Roh, Moo-hyun 2005: *Address on State Affairs Marking the Second Anniversary of His Inauguration*, 25 February.

55. Lee, Myung-bak 2008: "A Great People with New Dreams", *Address by President Lee Myung-bak on the 63rd Anniversary of National Liberation and the 60th Anniversary of the Founding of the Republic of Korea*, 15 August.
56. Lee, M.B. 2013.
57. Checkel, Jeffrey T. 2005: "International Institutions and Socialization in Europe: Introduction and Framework", *International Organization*, Vol. 59, No. 4, p. 804.
58. Lee, Myung-bak 2012b: *Address by President Lee on the 67th Anniversary of Liberation*, 15 August.
59. Secretary General of the United Nations 2001: "Major-General Jin Ha Hwang Appointed Force Commander of United Nations Peacekeeping Force in Cyprus", *SG/A/781*, 13 December. Available at: www.un.org/press/en/2001/sga781.doc.htm [Accessed 5 November 2018].
60. National Assembly Minutes 2013: *Standing Committee on Foreign Affairs Session 19-321-3*, p. 7.
61. Ibid.
62. Ha, J.S. 2015.
63. Ibid.
64. Permanent Mission of the Republic of Korea to the United Nations webpage: http://un.mofa.go.kr/english/am/un/bilateral/policy/activities/.
65. *Ministry of National Defense* 2012: p. 3.
66. National Assembly Minutes 2009: *Standing Committee on Foreign Affairs Session 18-283-4*, p. 2.
67. National Assembly Minutes 2011: *Standing Committee on Foreign Affairs Session 18-304-2*, p. 4.
68. Kim, Sung-Hwan 2011: *Keynote Speech by Mr. Kim Sung-Hwan, Minister of Foreign Affairs and Trade, at the 5th ROK-US Alliance International Conference*, 19 March.
69. Daniell, James 2013: "Sichuan 2008: A Disaster on an Immense Scale", *BBC News*, 9 May. Available at: www.bbc.com/news/science-environment-22398684 [Accessed 5 November 2018].
70. National Assembly Minutes, 2008: *Special Committee on Reviewing the Extension of the UNIFIL Mandate Session 18-276-1*, p. 7.
71. Fischer, Max 2013: "Japan and South Korea Can't Even Cooperate over Peacekeeping in South Sudan", *The Washington Post*, 26 December. Available at: www.washingtonpost.com/news/worldviews/wp/2013/12/26/japan-and-south-korea-cant-even-cooperate-over-peacekeeping-in-south-sudan/?utm_term=.edddfccef050 [Accessed 5 November 2018]; Ha, Eo-young and Seok, Jin-hwan 2013: "S. Korean Troops Borrow Japanese Ammunition in South Sudan", *The Hankyoreh*, 31 December. Available at: http://english.hani.co.kr/arti/english_edition/e_international/617788.html [Accessed 5 November 2018].
72. National Assembly Minutes, 2006: *Standing Committee on Foreign Affairs Session 17-262-17*.
73. National Assembly Minutes, 2006: *Plenary Session 17-263-1*.
74. Putnam, Robert D. 1988: "Diplomacy and Domestic Politics: The Logic of Two-Level Games", *International Organization*, Vol. 42, No. 3, pp. 427–460.
75. Yoon, Soon Gu 2016. "From Beneficiary to Benefactor: The Case of Korea", *Korean Institute for Defense Analysis, ROK Angle 142*, 24 October.
76. *Ministry of National Defense* 2008: p. 136.
77. Ibid.
78. Lee, Myung-bak 2012a: *Address by President Lee Myung-bak on the 93rd March First Independence Movement Day*, 1 March.
79. National Assembly Minutes 2014: *Standing Committee on Foreign Affairs Session 19-329-5*.
80. Lee, Jeongkyu 2014: "ROK's Contribution to World Peace", *Korean Institute for Defense Analysis, ROK Angle 114*, 22 December.

81. Ibid.
82. Hong, K.D. 2009: p. 173.
83. National Assembly Minutes 2009: *Plenary Session 18-283-1*, p. 8.
84. National Assembly Minutes 2009: *Plenary Session 18-285-1*, p. 18.
85. National Assembly Minutes 2009: *Plenary Session 18-283-1*, p. 8.
86. Ibid.
87. National Assembly Minutes 2009: *Plenary Session 18-285-1*, p. 18.
88. National Assembly Minutes 2006: *Plenary Session 17-263-1*, p. 50.
89. Ibid.: p. 51.
90. National Assembly Minutes 2009: *Plenary Session 18-283-1*, p. 10.
91. Ibid.
92. Ibid.: p. 11.
93. National Assembly Minutes 2012: *Standing Committee on Foreign Affairs 19-311-9*, p. 22.
94. Ibid.
95. National Assembly Minutes 2009: *Plenary Session 18-283-1*, p. 18.
96. National Assembly Minutes 2006: *Plenary Session 17-263-1*, p. 51.
97. Ibid.: p. 52.
98. National Assembly Minutes 2012: *Standing Committee on Foreign Affairs 19-311-9*, p. 22.
99. Lee, M.B. 2009a.
100. Lee, M.B. 2008.
101. Roh, Moo-hyun 2007: *Address by President Roh Moo-hyun on the 88th March First Independence*, 1 March.
102. National Assembly Minutes, 2009: *Plenary Session 18-285-1*, p. 18.
103. Ha, J.S. 2015.

References

Baek, Glenn 2013: "A Perspective on Korea's Participation in the Vietnam War", *The Asan Institute for Policy Studies Issue Brief 53*, 9 April. Available at: http://en.asaninst.org/contents/issue-brief-no-53-a-perspective-on-koreas-participation-in-the-vietnam-war/ [Accessed 5 November 2018].

Breuker, Remco E. 2009: "Korea's Forgotten War: Appropriating and Subverting the Vietnam War in Korean Popular Imaginings", *Korean Histories*, Vol. 1, No. 1, pp. 36–59.

Checkel, Jeffrey T. 2005: "International Institutions and Socialization in Europe: Introduction and Framework", *International Organization*, Vol. 59, No. 4, pp. 801–826.

Chung, Eun-sook 2010: "Korea's Law on UNPKO and Participation in International Peacekeeping Missions", *Korea Focus Current Issues and Policy, No. 166.* Available at: www.koreafocus.or.kr/design2/layout/content_print.asp?group_id=102934 [Accessed 5 November 2018].

Cumings, Bruce 2005: *Korea's Place in the Sun: A Modern History*, New York: W. W. Norton Company.

Daniell, James 2013: "Sichuan 2008: A Disaster on an Immense Scale", *BBC News*, 9 May. Available at: www.bbc.com/news/science-environment-22398684 [Accessed 5 November 2018].

Fischer, Max 2013: "Japan and South Korea Can't Even Cooperate over Peacekeeping in South Sudan", *The Washington Post*, 26 December. Available at: www.washingtonpost.com/news/worldviews/wp/2013/12/26/japan-and-south-korea-cant-even-cooperate-over-peacekeeping-in-south-sudan/?utm_term=.edddfccef050 [Accessed 5 November 2018].

Ha, Eo-young and Seok, Jin-hwan 2013: "S. Korean Troops Borrow Japanese Ammunition in South Sudan", *The Hankyoreh*, 31 December. Available at: http://english.hani.co.kr/arti/english_edition/e_international/617788.html [Accessed 5 November 2018].

Ha, Jong-Sik 2015: "ROK's Achievements through Its PKO Activities and Future Prospects", *Korean Institute for Defense Analysis, ROK Angle 122*, 29 May.

Heo, Uk and Roehrig, Terence 2014: *South Korea's Rise: Economic Development, Power, and Foreign Relations*, Cambridge: Cambridge University Press.

Hong, Kyudok 2009: "South Korean Experiences in Peacekeeping and Plan for the Future", *Advances in Military Sociology: Essays in Honor of Charles C. Moskos, Contributions to Conflict Management, Peace and Economics and Development*, Vol. 12A, pp. 173–187.

Howe, Brendan and Kondoch, Boris 2014: "Northeast Asian Perspectives on UN Peacekeeping: China, Japan, Korea", *Journal of International Peacekeeping*, Vol. 18, No. 3–4, pp. 133–153.

Hwang, Balbina Y. 2012: "Korea and PKO: Is Korea Contributing to Global Peace?", in: Scott A. Snyder (ed.), *Global Korea: South Korea's Contributions to International Security*, New York/Washington, DC: Council on Foreign Relations, pp. 13–27.

Ikenberry, G. John and Mo, Jongryn 2013: *The Rise of Korean Leadership, Emerging Powers and International Order*, New York: Palgrave Macmillan.

Kim, Bong-hyun 2013: *Statement by H.E. Ambassador Kim Bong-Hyun, Deputy Minister for Multilateral and Global Affairs, the Ministry of Foreign Affairs and Trade of the Republic of Korea, Open Debate on 'Peacekeeping: A Multidimensional Approach'*, 21 January.

Kim, Sung-Hwan 2011: *Keynote Speech by Mr. Kim Sung-Hwan, Minister of Foreign Affairs and Trade, at the 5th ROK-US Alliance International Conference*, 19 March.

Ko, Sangtu 2012: "Korea's Middle Power Activism and Peacekeeping Operations", *Asia Europe Journal*, Vol. 10, No. 4, pp. 287–299.

Lee, Jeongkyu 2014: "ROK's Contribution to World Peace", *Korean Institute for Defense Analysis, ROK Angle 114*, 22 December.

Lee, Myung-bak 2008: "A Great People with New Dreams", *Address by President Lee Myung-bak on the 63rd Anniversary of National Liberation and the 60th Anniversary of the Founding of the Republic of Korea*, 15 August.

Lee, Myung-bak 2009a: *New Year Policy Address to the Nation by President Lee Myung-bak*, 2 January.

Lee, Myung-bak 2009b: *Radio and Internet Address to the Nation by President Lee Myung-bak*, 30 November.

Lee, Myung-bak 2012a: *Address by President Lee Myung-bak on the 93rd March First Independence Movement Day*, 1 March.

Lee, Myung-bak 2012b: *Address by President Lee on the 67th Anniversary of Liberation*, 15 August.

Lee, Myung-bak 2013: *The Nation Brand Value Soaring: From Korea Discount to Korea Premium*, 21 January.

Lee, Shin-wha and Park, Joon Sung 2014: "Contributor Profile: South Korea, International Peace Institute Peacekeeping Database", *providingforpeacekeeping.org*, 13 June. Available at: https://s3.amazonaws.com/ipi-pdf-document-store/ppp-profiles/asia/ipi-pub-ppp-Korea.pdf [Accessed 5 November 2018].

Ministry of National Defense 2008: *Defense White Paper 2008*.

Ministry of National Defense 2010: *Defense White Paper 2010*.

Ministry of National Defense 2012: *Defense White Paper 2012.*
Ministry of National Defense 2014: *Defense White Paper 2014.*
Ministry of National Defense 2016: *Defense White Paper 2016.*
National Assembly Minutes 2006: *Plenary Session 17-263-1.*
National Assembly Minutes 2006: *Standing Committee on Foreign Affairs Session 17-262-17.*
National Assembly Minutes 2008: *Plenary Session 18-276-5.*
National Assembly Minutes 2008: *Special Committee on Reviewing the Extension of the UNIFIL Mandate Session 18-276-1.*
National Assembly Minutes 2009a: *Plenary Session 18-283-1.*
National Assembly Minutes 2009b: *Plenary Session 18-283-2.*
National Assembly Minutes 2009c: *Plenary Session 18-285-1.*
National Assembly Minutes 2009d: *Standing Committee on Foreign Affairs Session 18-283-4.*
National Assembly Minutes 2011: *Standing Committee on Foreign Affairs Session 18-304-2.*
National Assembly Minutes 2012: *Standing Committee on Foreign Affairs 19-311-9.*
National Assembly Minutes 2013a: *Plenary Session 19-320-4.*
National Assembly Minutes 2013b: *Standing Committee on Foreign Affairs Session 19-321-3.*
National Assembly Minutes 2014: *Standing Committee on Foreign Affairs Session 19-329-5.*
Park, Geun-hye 2015: *Remarks by the President at the New Year Press Conference*, 12 January.
Putnam, Robert D. 1988: "Diplomacy and Domestic Politics: The Logic of Two-Level Games", *International Organization*, Vol. 42, No. 3, pp. 427–460.
Roh, Moo-hyun 2004: *Address by President Roh Moo-hyun on the 59th Anniversary of National Liberation*, 15 August.
Roh, Moo-hyun 2005: *Address on State Affairs Marking the Second Anniversary of His Inauguration*, 25 February.
Roh, Moo-hyun 2006: *State of the Nation Address by President Roh Moo-hyun Presented to the 262nd Regular Session of the National Assembly*, 6 November.
Roh, Moo-hyun 2007: *Address by President Roh Moo-hyun on the 88th March First Independence*, 1 March.
Ruffa, Chiara 2014: "What Peacekeepers Think and Do: An Exploratory Study of French, Ghanaian, Italian, and South Korean Armies in the United Nations Interim Force in Lebanon", *Armed Forces & Society*, Vol. 40, No. 2, pp. 199–225.
Secretary General of the United Nations 2001: "Major-General Jin Ha Hwang Appointed Force Commander of United Nations Peacekeeping Force in Cyprus", *SG/A/781*, 13 December. Available at: www.un.org/press/en/2001/sga781.doc.htm [Accessed 5 November 2018].
Sesay, Amadu 1997: "Dividends on Investments in History: Korea and UN Peace-Keeping and Peace-Building Operations", *Korea Observer*, Vol. 28, No. 2, pp. 181–217.
Shin, Kak-soo 2010: *Keynote Address by H.E. Dr. Shin Kak-soo, Vice Minister of Foreign Affairs and Trade, at the Luncheon Hosted by the UN Association of the Republic of Korea on UN Day*, 24 October.
Song, Min-sun 2007b: *Foreign Minister Song Min-soon's Interview with Arirang TV on 'Diplomacy Lounge'*, 29 December.

Yoon, Soon Gu 2016: "From Beneficiary to Benefactor: The Case of Korea", *Korean Institute for Defense Analysis, ROK Angle 142*, 24 October.
Yu, Myung-hwan 2008: *Speech by H.E. Yu Myung-hwan, Minister of Foreign Affairs and Trade, at the Luncheon Hosted by the UN Association of the Republic of Korea on the 63rd Anniversary of UN Day*, 20 October.
Yu, Myung-hwan 2009: "Diplomacy at Forefront to Realize Global Korea", *The Korea Times*, 1 March.

6 South Korea and Climate Diplomacy

The following chapter is the second case study, South Korean international climate diplomacy, which also has been identified as one of the key areas of South Korean global foreign policy ambition in the respective literature.[1] The first section briefly introduces the history, as well as the main literature, on South Korean climate diplomacy, followed by the results from the empirical analysis. The focus hereby lies on role conceptions and exhibited key themes put forward and contested by different actors in the South Korean climate diplomacy discourse of the last decade. The main finding is that while the main role conception of being a "bridge" has been stable and dominant over the observation period, the inherent tension between Korea's global role model aspirations and its actual brown growth development track was only resolved through the focus on international leadership and the promotion of a green growth strategy. Only then was Korea able to enact an internationally recognized role that fitted its own status and self-image. In consequence, the aspect of the "rising Korea" self-image about being a country still in development and thus in need of prioritizing economic growth over climate change concerns made way for a green growth vision that still focused on economic opportunities, but this time in low-carbon future-growth industries and markets. However, with a changing international context and a global momentum with the Paris Agreement, South Korea's main goal then was to use the established green growth platforms and climate diplomatic channels to maintain and further expand the country's network and influence as a green leader and climate hub.

An Introduction to South Korean Climate Diplomacy

The starting point for international climate change diplomacy was the Rio Earth Summit in 1992, where the UNFCC treaty aiming at "the stabilization of greenhouse gas concentrations in the atmosphere at a level that would prevent dangerous anthropogenic interference with the climate system"[2] (Art. 2) was signed. The parties to the convention agreed to work toward this goal "on the basis of equity and in accordance with their common but differentiated responsibilities and respective capabilities"[3] with developed countries taking the lead (Art.3(1)). Since 1995 the parties have met annually at the Conferences of the Parties (COP) to assess progress and debate further measures in fighting climate change, such as the Kyoto Protocol in 1997 or the Paris Agreement in 2015.

In the early 1990s, South Korea was still categorized as non–Annex I country, meaning a developing nation under the UN Framework Convention on Climate Change (UNFCCC), and most Korean governments since then have tried to retain this status while participating in an increasing number of environmental negotiations but ultimately "maintain[ing] a passive stance as an observer of international climate change issues."[4] The major concern of the administrations back then was to ensure continuing growth for the economy amid increasing global competition and especially in the aftermath of the 1998 Asian Crisis or IMF crisis, as it is known in Korea. Yoon argues that Seoul's attitude toward climate change diplomacy and global environmental politics on the whole still "reflect[ed] its prioritization of its economic interests in its foreign policy."[5] Furthermore, as Watson points out, under the authoritarian strongmen of previous decades, environmental concerns were seen as dangerous subversion, because economic development has been the key for South Korea's main national security strategy toward North Korea.[6]

At the same time, environmental concerns were widespread among the South Korean public, as the decades of compressed modernization and development also came at the price of environmental depletion and pollution. With regard to greenhouse gas emissions, Peritore wrote that, in 1998, 77 percent of Koreans were "in favour of major action on climate change issues"[7] and that the international community increasingly demanded Seoul's "participat[ion] in global efforts for environmental protection in ways that are commensurate with its economic capacity and emissions volumes."[8] Prior to 2008, however, South Korea's involvement in global climate change negotiations remained "practically non-existent,"[9] until President Lee Myung-bak's administration announced the so-called Low Carbon Green Growth strategy in 2009.

President Lee's green growth strategy was an ambitious plan to transform the country's economy which was based on "brown growth" industrial development through a Green New Deal, making innovative green industries the future-growth engines. In addition to this domestic dimension, "green growth" was also expected to enable South Korea to enact "multiple leadership roles"[10] in the international sphere. The explicitly stated strategic objective was to become a global "role model"[11] for green growth, which was sought to be achieved through performing a bridging role in international climate change negotiations, the promise to allocate significant ODA funds for green growth projects, and the establishment of the GGGI in order to become a global hub for green growth know-how.[12]

Generally, South Korea's diplomatic green growth offensive is seen as exemplary commitment and a success, for example, by the World Bank[13] and the OECD.[14] Korea not only established the GGGI as an international organization in 2012 but also won the hosting right for the UNFCCC Green Climate Fund (GCF) in 2012. Accordingly, Ikenberry and Mo attribute to South

Korea the roles of "agenda setter" for promoting green growth internationally in the OECD and UNEP, "mediator" for facilitating the UNFCCC registry for Nationally Appropriate Mitigation Actions (NAMAs) involving developing countries at COP 15 in Copenhagen 2009, and "bridge" for hosting the GCF. Due to the successful international promotion of "green growth," Blaxekjær concludes that South Korea has established itself internationally as a major actor in international environmental politics and hence categorizes Seoul as a "green middle power."[15] Ikenberry and Mo even see Seoul's green growth efforts as "driven by liberal international values in terms of substance as well as formality."[16] Others[17] speak of South Korea as a middle power more generally in international climate diplomacy, but there is also some criticism about the South Korean projection of a green image, especially in the context of the discrepancy between international aspirations and domestic environmental performance.[18] Here the green growth paradigm is seen as little more than another top-down,[19] developmental, state-driven economic stimulus package,[20] involving polarizing measures such as the large-scale Four River Restoration project as well as the promotion of nuclear power.[21] Furthermore, while generally agreeing with the need for a greener developmental model, the South Korean public has not been really supportive of global leadership claims: in the 2013 Public Opinion Survey on the Green Growth Policy undertaken by Lee Myung-bak's Presidential Committee on Green Growth (PCGG), no more than 2.8 percent of the respondents gave a priority to the enactment of global leadership. Over half of the respondents (54.6 percent), however, were primarily concerned with renewable energy sources.[22]

In particular, the analysis is focused on the Framework Act on Low Carbon and Green Growth in 2009, the support for the establishment of the GGGI in 2012, and the bid to host the GCF in 2012, as well as the ratification of the Paris Agreement in 2016. Because the wider political discourse consists of the government, in this case the president and the Ministries of Foreign Affairs and Ministry of Environment, as well as the political opposition, this approach foregrounds domestic contestation (especially in the National Assembly), as well as changes over time. At the center of the analysis are questions about which specific role conceptions and self-images have been used by different political actors to justify the climate diplomacy and green growth legislation.

Role Conceptions, Key Themes, and Significant Others

There are a number of retrieved role conceptions that have been employed in the South Korean discourse on climate diplomacy from 2005 to 2016 throughout the political spectrum: the consistently influential "bridge between developed and developing world," the standout "green leader," and two marginal, rivaling responses to the latter, the traditional "non–Annex I country" and the more recent network-focused "climate hub."

In contrast to the political debate in the US, for example, throughout the whole observation period there has been a political consensus that climate change and global warming are human-made through increasing greenhouse gas emissions. Furthermore, neither the fact that South Korea is a major carbon emitter nor that the country has a high dependence on energy imports was politically questioned or challenged. The main political issues with regard to international climate diplomacy were, first, how to balance economic growth with the legitimate need for lower carbon emissions and, second, whether Seoul should pursue a global leadership strategy in international climate talks. The stable, main role conception underlying South Korea's climate diplomacy engagement was that of a "bridge between developing and developed world," although with regard to the concrete meaning of this rather wide trope there have been two progressions. On one hand, President Lee Myung-bak's green growth promotion drastically redefined South Korea's bridge role conception in climate diplomacy while putting Seoul under some pressure to live up to these newly raised expectations about Korea being a "green leader" around the time of the Paris Agreement.

This can also be illustrated with the voting figures on the controversial "Framework Act on Low Carbon and Green Growth" from 2009, which was a key piece of legislation of the Lee Myung-bak administration, and the ratification of the Paris Agreement in 2016. As one can see in Table 6.1, South Korea's commitment to a proactive diplomatic stance on cutting carbon emissions became less controversial over time. The Framework Act even saw "no" votes from all parties present in parliament at that time. A more detailed description of the political debates and various arguments is given in the following sections.

A key factor for this development is to be found in the change in the international context. Under the Kyoto framework, international negotiations were initially locked in a diplomatic stalemate over the unequal responsibilities to cut carbon emissions between developing and developed countries. In the lead-up to the successful COP 21 in 2015, however, a deal struck between China and the US in 2014 changed the momentum and enabled the signing of the Paris Agreement.

The Main Role Conception: "Bridge Between Worlds"

The key trope used in the South Korean domestic discourse on international climate diplomacy over the whole observation period from 2005 to 2016 is that of a "bridge between developing and developed countries." This is also similar

Table 6.1 Voting Shares of Major Climate Diplomacy Legislation, National Assembly of the ROK, 2009–2016

Framework Act on Low Carbon and Green Growth, 29 December 2009	**Ratification of the Paris Agreement, 3 November 2016**
(Source: National Assembly Minutes, 2009: *Plenary Session 18–285–1.*)	(National Assembly Minutes, 2016: *Plenary Session 20–346–11.*)
Yes votes: 140	Yes votes: 255
No votes: 40	No votes: 0
Abstentions: 10	Abstentions: 6

to the debate on overseas developmental assistance where the South Korean elites like to stress their country's developmental success story, the "miracle on the Han River," which uniquely qualifies South Korea to understand and speak on behalf of the developing countries' interests. All three observed administrations consistently argued for a "bridging" role of South Korea in international affairs and especially climate diplomacy.

For President Roh Moo-hyun himself, global climate diplomacy was not a high priority, but his general vision for Korea saw the country as a central hub or node for regional relations. Roh's idea about Korea's bridging role was decidedly regional in scope while he focused on the geographical centrality of the Korean Peninsula in Northeast Asia:

> The Korean Peninsula is located at the heart of the region. It is a big bridge linking China and Japan, the continent and the ocean. Such a geopolitical characteristic often caused pain for us in the past. Today, however, this same feature is offering us an opportunity. Indeed, it demands that we play a pivotal role in the Age of Northeast Asia in the 21st century.[23]

His foreign minister, Song Min-soon, however, made clear that Korea as "an OECD member country and a major greenhouse gas emitter"[24] planned to enact this bridge role also in international efforts related to climate change:

> As a newly industrializing country standing between developing and developed nations, Korea will work together with the international community in devising innovative solutions to climate change, in an effort to bridge the gap between developed and developing countries by forming a strong partnership between the two.[25]

Song further calls on the international community to work together toward a "more flexible" post-Kyoto framework that succeeds in accommodating the interests of developed countries skeptical of the Kyoto Protocol, for example, the US and Australia, as well as fast-growing economies of developing countries, such as China or India. This is indicative of the conundrum South Korea saw itself in within the Kyoto regime: on one hand, the country is eager to take up international responsibilities as an OECD country and does not dispute its high share in greenhouse gas emissions, but, on the other hand, it has been categorized as a non–Annex I country and intends to safeguard its status as a "developing country," exempt from binding restrictions on CO_2 emissions.

Against this backdrop, the role of "bridge" appears to be a fitting solution to the climate diplomatic dilemma. Through evoking South Korea's recent developmental history, the government is able to carve out a special position as a "bridge" to project the image of a responsible nation and to match international expectations while avoiding strict regulations for domestic businesses. Similarly, in 2009, under the presidency of the conservative Lee Myung-bak, Kim Hyeong-guk, the chairman of the newly established PCGG, describes South Korea's role in international climate diplomacy as "bridge" while stressing that Korea has no historical responsibility for climate change:

> As a member of the OECD and as an emitter of greenhouse gases, we are committed to participating in global climate change efforts that correspond to our national strength. As a non-Annex I country, we plan to voluntarily announce country-specific reduction targets similar to those of developed countries. [. . .] However, since Korea has different historical responsibilities for climate change than developed countries, we also seek to pursue a different approach than developed countries. Furthermore, we will enhance the image of the country through constructive negotiation proposals and we will build trust in the international community and act as a bridge for developing countries.[26]

Korea's role here is conceptualized as "mediator" or "honest broker," particularly for developing countries' interests. However, there is a major problem with the credibility and plausibility of this narrative as it is. South Korea's economic development was achieved through CO_2-intensive industrialization or brown growth, and thus, the South Korean developmental experience can hardly serve as a good example of sustainable development. This problem is only solved through President Lee's international promotion of green growth as a new developmental paradigm from 2008 onward.

The Complementary Role Conception: "Green Leader"

Despite the legitimate criticism that the concept of green growth prioritizes economic growth over sustainability and environmental concern, Seoul's global promotion of its green growth profile fixes the credibility problem faced by a bridge narrative based on a brown growth development experience. The promotion of green growth domestically and internationally was—together with the Global Korea national security strategy—a trademark policy of President Lee Myung-bak. Even though the Seoul Initiative Network on Green Growth (SINGG) had already been established under President Roh Moo-hyun in 2005, it was only made a major policy under President Lee Myung-bak.[27] The extent to which President Lee invested his own personal reputation for the green growth agenda explains also his standout discursive position in the climate diplomacy discourse. Blaxekjær even concludes that green growth became "synonymous with President Lee"[28] and reports how the OECD called Lee the "father of green growth."[29]

For the Lee administration, green growth was the fitting next step in South Korea's development and its roles as a bridge, a global leader, and—after its admission to the G20—a role model can be seen in the following quote by President Lee from 2012:

> Korea is the only country that rose to a member of the G20 from a poverty-stricken nation in a matter of a generation. Many developing countries are

> benchmarking Korea as a role model. Furthermore, our nation is in the midst of successfully overcoming environmental destruction caused by fast growth and development. Thanks to these experiences, the Republic of Korea has exercised great persuasive power in spreading the value of green growth to developing countries that are experiencing rapid growth.[30]

Here, Korea has two experiences to offer: first, the success story of the rise to a G20 nation due to economic development and, second, the recent experience as a green growth pioneer. Both together support Seoul's claim to act as a "bridge between developed and developing world," while the country still only commits to voluntary and not internationally binding measures, according to its non–Annex I status. When announcing voluntary reductions of carbon emissions in 2010, President Lee alluded to moral obligations instead:

> Last year, Korea announced that it will reduce its carbon emissions 30% of BAU by 2020. Korea is a non-annex I country so we have no legal obligations but we understand we have moral obligations. This is why we voluntarily decided to abide without any conditions [to] the highest standard set by the international community.[31]

The central balancing act between economic interests and international climate commitments was prominent during the debate about the bill for a "Framework Act on Low Carbon and Green Growth" in the National Assembly's Special Committee on Measures against Climate Change on 14 April 2009. The oppositional DP lawmaker U Je-chang, for example, raised the question of whether the South Korean economy's industrial base and its competitiveness vis-à-vis other developing countries such as China and India would be unnecessarily damaged by the government's voluntary low carbon measures.[32] Representative Kim Nak-seong, from the small conservative Liberty Forward Party, also challenged the government's assessment of an international trend toward more carbon trading initiatives and suggested a more cautious "wait and see" approach before committing to any climate action that may hinder Korea's international competitiveness.[33] Because of the international stalemate between Annex I and non–Annex I countries under the Kyoto regime, namely, between the US and China, these lawmakers did not see an international context given where Korea should risk damage to its international competitiveness by constraining its economy with legislating carbon cuts. The conservative GNP lawmaker Jeong Du-eon also questioned Korea's responsibilities with regard to climate change in general: "We actually do not have a historical responsibility. That's why we are being treated as a developing country and not as an OECD country."[34] For others, for example, DP lawmaker Kim Jae-yun, the government's green growth strategy either was lofty and ambitious talk by the president with little benefit for the well-being of the people[35] or was seen as not really all that "green" for a lack of focus on sustainability, as in the case of Creative Korea Party's opposition lawmaker Yu Weon-il.[36]

To summarize this briefly, the main objections to President Lee's green growth agenda were that it was either unnecessarily ambitious and risky for Korea's economic competitiveness or not really sustainable and "green" enough. Both positions can be read as direct responses to the "green leadership" role aspiration, even though both lost influence and remained rather marginal throughout the observation period. Still, the first objection can be seen as the traditional "non–Annex I country" role conception that prioritized economic development over environmental concerns and was held by South Korea during the 1990s and early 2000s,[37] whereas for other marginalized voices the "green leadership" was not green or environmentally friendly enough.

Moreover, there were two major domestic aspects of the bill that faced particular opposition in parliament. First, parliamentarians successfully stopped the inclusion of the promotion and export of nuclear power in the law, and second, oppositional lawmakers expressed their concern over one of President Lee's main campaign pledges, the construction of a Grand Channel through the country from coast to coast. The latter project, however, later became a reality in a watered-down version as the Four Major River Restoration project, despite significant resistance from the political opposition, the media, and the public. Notably, both policies were framed as measures to fight climate change or to adapt to it. In the case of nuclear power, President Lee argued that "[o]ur development strategy aims to make Korea a leader in green growth by implementing proactive and pre-emptive policies that will tackle climate change. Exporting nuclear power plants is one good example of such a strategy."[38] And with regard to the Four River Project, the president had leveraged general international support for the project:

> Korea's climate change policy received high praise at the Conference of the Parties of the UN Framework Convention on Climate Change held in Cancun, Mexico, at the end of last year, and was ranked as the second best out of 57 countries. The Four Major Rivers Restoration Project is also highly recognized in the international community. The UN Environment Program announced that Korea's Four Rivers Project is an exemplary model of environment-friendly endeavors as well as an efficient means of coping with climate change.[39]

Other investments planned were in renewables, research, and development as well as "green" cars and infrastructure, in order to make Korea a "leading powerhouse in the green technology market."[40] The declared goal was to boost domestic growth while making the country more energy-independent. President Lee's call for the transformation of the national economy was presented to the country with a "we have caught up with the world before" spirit. Take, for example, these quotes from his address on the 63rd anniversary of national liberation and the 60th anniversary of the founding of the ROK on 15 August 2008:

> Green growth will enable a Miracle on the Korean Peninsula to succeed the Miracle on the Han River. When the Republic first manufactured its own vehicles, the technology gap with the advanced countries amounted to at least 50 years. As far as semiconductors are concerned, it was more than 20 years behind. However, the Republic grew into a technology powerhouse, which ranks first in terms of producing semiconductors and ships, and fifth in automobiles. If we make up our minds before others and take action, we will be able to lead green growth and take the initiative in a new civilization. To do this, I will make sure that the country comes up with new green growth engines for the next generation to use for 10 to 20 years.
>
> [. . .]
>
> If Korea makes an audacious and swift move just as it did to advance its information capabilities to make up for belated industrialization, the country will undoubtedly be reborn as a green power.[41]

President Lee framed the moment as an opportunity for Korea to put its economy ahead of other international competitors, if the country follows his plan for yet another rebirth as a "green power." Using the historical precedent of the miraculous developmental modernization undertaken in the last decades, the President tried to build support for his contested domestic agenda.

In another way of justifying his "green growth" agenda, Lee regularly references proudly the international acknowledgment and recognition for Korea's "green growth" agenda. In his retelling, the strategy paid off, and the world recognized the global contributions by Korea, which, in turn, proves the country's (and thus his own) successful leadership, as the following quotes from speeches mostly from Liberation Day addresses in 2009, 2011, and 2012 illustrate:

> We have already applied the idea of green growth to our economy and see the results that what is better for the economy has proven to be better for the environment and vice versa. Last June, the OECD officially adopted this Korean initiative as an official ministerial declaration. The importance of this method is recognized by the world.[42]
>
> [C]limate change constitutes a task that requires the joint endeavors of the global community without any further delay. However, the differences of opinions between advanced countries and newly emerging economies, including China and India, still remain unabated. In order to address this problem particularly, I put forward a proposal to set up a working group to take concrete measures. The co-chair of the summit, US President Barack Obama, accepted the proposal on the spot, saying that it was a very effective and practical suggestion.[43]
>
> Our vision of green growth is the first time in our history where a global issue is recognized as our own. It is also pivotal to our mission of becoming a greater Republic of Korea.[44]

> As part of the efforts to respond to climate change, Korea worked together with many countries around the world and established the Global Green Growth Institute.
>
> [. . .] Through the Institute, the first international organization established at Korea's initiative, green growth has become an enduring goal for the international community.
>
> Over the years, the Republic of Korea has grown into a greater country, and its global leadership has been further cemented.[45]

Also in 2012, he left no doubt about his own personal share in this diplomatic success:

> In my 2008 Liberation Day remarks, I declared green growth, which is aimed at striking a balance between the economy and the environment, as the new vision for national progress. This vision has now become a mega trend embraced by many countries around the world. As a result, the term "green growth" emanating from Korea is now a part of the common language of the world.[46]

On other occasions, President Lee cited the emotional praise, tears of joy, and gratitude expressed toward him by leaders from developing nations or his expressing surprise over the positive effects Korea's and his own leadership have had on the world. Lee uses these anecdotes as testaments for the appropriateness of his strategy for international green growth promotion. For example, when South Korea won the hosting rights for the UN GCF in 2012, President Lee commented proudly,

> It is the first time for an Asian country, let alone Korea, to become the host of such a significant international organization. Last year when Korea unveiled its intention to host the organization, everyone thought our chances were slim. Germany, one of the competing rivals, is the most renowned green nation in the world. [. . .] For these reasons, Korea was regarded as the underdog in the competition.
>
> The news that Korea beat out rivals to host the organization despite such disadvantages came as a surprise upset to the international media. One head of state said to me, "The Republic of Korea is able and qualified to contribute to the international community. I supported Korea because it has sufficient capabilities."[47]
>
> At the GCF council meeting on October 17, a South African delegate praised Korea with emotion saying that he wants to share Korea's experiences in national development and green growth. By playing the role of a bridge between the developing and developed worlds, our country can contribute to the international community and find our own future there.[48]
>
> The leaders who were most impressed about the summit results were those from poor African nations. When I met them, they grabbed my hands

> and expressed their genuine thanks with tears filling their eyes. [. . .] The reason we were able to rouse gratefulness in these nations is that we can read their mind and understand their situations quite well.[49]

After establishing the GGGI as an international organization and securing the hosting rights for the GCF in Incheon, it is fair to assess South Korea's role-playing as a leading power in climate diplomacy and as a green pioneer. Through President Lee's promotion of green growth, South Korea established itself as a credible "bridge between developing and developed countries" in international climate diplomacy. This strategic move was helped, however, by a stalemate in climate diplomacy after the failure of the COP 15 in 2009 and a lack of effective leadership toward a post-Kyoto consensus regarding binding levels for CO_2 emissions in the aftermath. But with a rapprochement between the US and China in a deal to cut CO_2 emissions in 2014,[50] this diplomatic environment changed and led to the Paris Agreement in 2015. This changing international context also put South Korea's credibility as a "green leader" to the test, as will be shown in the following section.

International Context and Expectations: "Green Race"

Throughout the whole observation period, international expectations were a main issue in the South Korean climate diplomacy discourse. At the core of this was the awareness that Korea's official status as a developing country under the regulations of the Kyoto Protocol was only ever a temporary exemption from legally binding measures as the country was rising through the ranks of the international community. As Yun Jong-soo, a climate policy officer at the Ministry of Environment, put it in 2009 during a public hearing on climate change measures and green growth-related regulation, "it is doubtful whether Korea can remain a developing country while being in the OECD for 13 years."[51] Already in 2005 Deputy Minister of Environment Park Seon-sook had raised that point that even though Korea has no legal obligations, it has the fastest growing rate of carbon emissions among OECD countries[52] and even the deputy minister from the Ministry of Commerce, Industry, and Energy, Lee Weon-geol, conceded that "Korea is expected to play a responsible role corresponding with its status in the international society."[53]

While Korea was increasingly expected to pull its weight as an OECD country and major CO_2 emitter, in addition policy makers in Seoul saw a growing challenge in the gaining international momentum in favor of further measures to tackle greenhouse gas emissions. Yun Jong-soo, the climate policy officer at the Ministry of Environment, for example, warned of a worldwide "green race"[54] where plans for emission trading schemes by the European Union (EU) and the US would lead to "a situation where other countries cannot help but join this very rapidly."[55] Representative Lee In-gi, from the ruling conservative GNP, similarly argued for "a need to push forward, as advanced countries' climate change legislation is in full swing."[56] When making the case for the "Framework Act on Low Carbon and Green

Growth" in 2009, Prime Minister Chung Tae-shin referred to the international momentum and dramatically exclaimed, "Now this is not a question of our choice, it is a matter of necessity."[57] There was even a sense of risking falling even further behind "advanced nations"[58] if Korea was to miss this development, as expressed by GNP lawmaker Yu Gi-jun ahead of the (ultimately unsuccessful) COP 15 in Copenhagen: "It is clear that the Copenhagen meeting will greatly strengthen international obligations to reduce greenhouse gases and in such a situation Korea will be bound to participate in the efforts of the international community."[59] President Lee Myung-bak too referred to this international momentum and the global financial crisis since 2008 when he made the case for Korea to seize the opportunity and put itself at the top of the alleged international trend toward a greener economy and diplomacy. In 2011, for example, President Lee addressed the nation with the following words:

> During the 1970s, we experienced oil shocks twice. In 1997, we suffered a foreign exchange crisis. In 2008, there was a global financial crisis. But the nation has coped with all these and other hardships wisely. In fact, Korea is known internationally as a country that overcomes major crises in the most exemplary fashion.
>
> Whenever the nation is faced with major difficulties, I am reminded of the great Korean people with pride and gratitude for what they have overcome. I believe in the potential of the Republic of Korea that becomes even stronger in times of crisis.[60]

His successor, President Park Geun-hye, also repeatedly made the case for decisive climate action as an opportunity rather than a crisis. At the opening ceremony for the GCF in Incheon, President Park said,

> When viewed with an eye to action, the challenge posed by climate change is indeed an opportunity—to redefine our values, generate new markets and create new jobs. I am confident that we can combat climate change in more efficient and environmentally friendly ways with the help of science and technology.[61]

Furthermore, in her 2014 New Year address, she reaffirmed the leadership ambition that President Lee had expressed: "We too must step into the currents of change, not only joining but also taking the lead."[62] In many ways her administration's stance on climate diplomacy was a continuation of President Lee Myung-bak's policies, exhibiting the same complementary role conceptions of "bridge" and "green leader." For example, in her speech in front of the General Assembly of the UN in 2014, President Park promised that "[t]he Republic of Korea is ready to play a bridging role between developed and developing countries as the post-2015 development goals are being set, by harnessing our unique historical experience."[63] The following statement by her foreign minister, Yun Byung-se, also illustrates the continuing commitment to the climate diplomacy pursued by President Lee, focusing on the GCF and GGGI:

> To maintain its leading position in the field of climate change and to actively participate as a party in the follow-up negotiations for the implementation of the Paris Agreement, Korea has also been supporting the Green Climate Fund (GCF) and the Global Green Growth Research Institute (GGGI).[64]

Moreover, the parliamentary debate around the ratification of the Paris Agreement in 2016 is insightful especially with regard to the "green leader" role conception, as well as the sense of a "green race." The ratification of the agreement found resounding support both from the government as well as in the National Assembly. In the end, it was accepted with 254 "yes" votes out of 261 cast votes, with only seven abstentions (six from the center-left Minjoo Party and one from the centrist People's Party) and no "no" votes.[65] Whereas the government made the case for the ratification of the treaty with reference to the international consensus and the ambition to remain a leading green power in international climate diplomacy, conservatives, and progressives alike criticized the government for its lack of further concrete policy plans in order to cut CO_2 emissions. For example, in the discussion over the proposal in the Foreign Affairs Standing Committee on 28 October 2016, the chairman Lee Jong-Hu argued that "[i]n order to lead the discussion of the international community on the future climate change response, it is considered necessary to approve the ratification."[66] Furthermore, he referred to Korea's "international responsibility as a greenhouse gas emitter country."[67] Deputy Minister of Foreign Affairs Cho Tae-yeol also pointed out that because India, China, and the EU were all quickly progressing with the approval of the treaty, "we would be too lagging behind if we do not ratify [the Paris Agreement] quickly."[68]

In terms of criticism, Minjoo Party Representative U Weon-sik made clear that the Paris ratification is only part of what the government should do about fighting climate change. For him, the government's measures were not green enough, and so he referred to the international NGO Climate Action Tracker, which assessed South Korea's response to climate change as "insufficient." In line with earlier calls for a greener leadership role, he closed by demanding more action: "Even if our Parliament passes the ratification today, the government will have to propose additional reduction targets that match our status in the international community, so that we do not pass on the burden to future generations."[69] Despite the fact that his conservative colleague from the ruling New Frontier Party, Hong Il-pyo, also evoked the fate of future generations, Hong's emphasis lay on the "green growth" opportunities:

> Because developing and developed countries are participating this time in the Paris Agreement, it is a good opportunity for us in our middle-position

> and with our technological strength to take the lead in the fourth industrial revolution and the new energy market paradigm.[70]

Notable here is the absence of the traditional "non–Annex I country" role conception that tries to allow only as little environmental regulation as possible, for the sake of economic growth and international competitiveness. The main reason for this lies in the change of international context from a stalemate between developing and developed countries under the Kyoto regime, to a "green race" after the US–China deal on carbon emission in 2014 cuts as well as the Paris Agreement of 2015. The significant Others in South Korea's climate diplomacy are thus first and foremost the US, whose domestic climate change debate had been regularly discussed in the National Assembly, and international economic competitors, most important, China and Japan.

Self-Images

After identifying the main role conceptions and key themes in the South Korean wider foreign policy discourse on international climate diplomacy, the final step of the analysis consists of a close, intertextual, and interpretative reading of the utilized self-images, as well as their commonalities and differences. Through these pixelated snapshots of self-identification practices, understood as specific linkages between role conceptions and self-images, one can, in a next step, acquire a more comprehensive understanding of the underlying identity narratives informing and being informed by the global foreign policy actor South Korea.

The main self-image prevalent in the climate diplomacy discourse is that of a "rising Korea" with the country being at the brink of becoming an advanced nation or a country "in-between worlds." It is at the core of the "bridge" as well as the "green leader" role conception. Central to this self-image is the narrative about Korea's successful development, the "miracle from the Han River," as well as the transformation from "a beneficiary to a benefactor." Here, Korea has arrived, traveling from the periphery to the center, finally gaining a seat at the table. The country is aware of its rising responsibilities and is generally eager to live up to them. With President Lee's green leadership ambition, this self-understanding gets a redefinition and undergoes a further transformation. Here, Korea still pursues a role as a "bridge between developing and developed world" but not just because of its development history and risen status but also because of its potential and agenda as a "leading Korea."

The "green leader" role conception highlights the belief that the Korean nation has always thrived in crises and is now eager to teach this unique experience to the world. President Lee, for example, repeatedly urged the Korean nation to see climate change as "the most fundamental and colossal threat humankind has ever experienced"[71] and as an opportunity to adapt quickly and close up to more advanced countries, not simply through economic development or military might but also as agenda setting and norm entrepreneurship "for humanity":[72]

> Economic might and military power are still important. But, these alone will not make a country a global leader. What we need now is a vision that can be accepted and shared by the world. In this regard, Korea's "Low Carbon Green Growth" vision, which pursues sustainable development while responding to climate change, is widely recognized by the international community as a common vision for the future.[73]

Thus, the related self-image can be called "leading Korea." Already at the very beginning of his term, in one of his first major speeches, the address on the 63rd anniversary of national liberation and the 60th anniversary of the founding of the ROK on 15 August 2008, President Lee presented his leadership vision explicitly as a transformative one, aiming at fulfilling his "dream of an advanced nation."[74] He also spoke of being "reborn as a green power"[75] and the chairman of the PCGG, Kim Hyeong-guk, argued for the "Framework Act on Low Carbon and Green Growth" at some point to be in line with a "global morality of advanced countries."[76]

Furthermore, President Lee also promised that through his promotion of green growth internationally, Korea would be able to punch above its weight and carve out a new future path for development as a major Asian nation. In his rationale, he also made use of the trope of an "underdog," showing how Korea's potential had been underestimated internationally, for example, in the context of winning the GCF hosting rights:

> It is the first time for an Asian country, let alone Korea, to become the host of such a significant international organization. Last year when Korea unveiled its intention to host the organization, everyone thought our chances were slim. Germany, one of the competing rivals, is the most renowned green nation in the world. Worse yet, the Interim Secretariat of the Green Climate Fund is presently located in Bonn, Germany. For these reasons, Korea was regarded as the underdog in the competition. [. . .] In this connection, the fact that Korea was chosen as the host country of the Green Climate Fund carries added significance indeed.[77]

Also, the successful establishment of green growth institutions in international climate diplomacy, an issue area that is inherently about cooperation,[78] was a welcome opportunity to build further diplomatic ties and networks, making Korea a well-connected and highly regarded networking power or climate hub. Take, for example, this quote by President Lee from 2009:

> But I do not think we can cope successfully with issues that affect all mankind, such as climate change, with efforts by a small number of countries alone. I believe that only when all countries join efforts can they create

> a green big bang, a new world order, under which economic growth can be pursued in harmony with environmental protection. [. . .] I stress once again that Korea will continue to lead green growth by efficiently coping with climate change and through the formation of the East Asia Climate Partnership with neighboring countries.[79]

The establishment of the GGGI was also argued for in making the case for global partnerships beyond climate diplomacy, especially with Denmark, which later co-founded the institute with Korea. See, for example, these quotes from President Lee:

> For green growth to be sustainable, I believe global partnership is vital. No one living on this planet can escape from the devastation that is brought on by climate change. Therefore, finding a solution is our collective responsibility. As part of our contribution to global partnership, Korea proposed setting up the Global Green Growth Institute.[80]
>
> Cooperation with Denmark, which boasts the best green technology in the world, will serve as a great opportunity to open a new chapter in Korea's economic progress. The Danish economy substantiates the fact that green growth industries, as future growth engines, are conducive to creating new job opportunities. Between 2000 and 2008, the export of Denmark's environment-friendly technologies and goods increased by two and a half times, thereby accounting for 11 percent of its total exports.[81]

His successor, President Park Geun-hye, also utilized the green growth platform to further Korea's diplomatic influence, especially in Africa:

> In responding to climate change issues such as land degradation, water shortage and food scarcity that confront many African states, Korea plans to take joint action with Africa through the Green Climate Fund and the Global Green Growth Institute based in Korea.[82]

In the same way, oppositional Democratic United Party lawmaker Hong Yeong-pyo was expecting that the opportunity to host the GCF would "improve the national brand image and make Korea a climate change hub."[83] Here another aspect of the "leading Korea" self-image becomes apparent: as the country is seeking ties around the world, in order to expand and secure its newfound influence and status as a leading nation in green growth, it has to live up to the international expectations and the context of a "green race." Given the high level of domestic support with the "green leader" role conception, the self-image as a "model country" that is taking the lead on climate change issues meant that Korea's appropriate action with regard to the Paris Agreement was to not "fall behind" and sign up to the global leadership or "green race" unfolding. The role conception of "bridge between worlds" remained the broader frame of Korea's engagement with the world, but it was only meaningfully manifested through the clear adherence to "green leadership." Or, put differently, without acting as a green leader, Korea would have trouble upholding the much-wider role identity of a "bridge," externally and internally.

This is in stark contrast to the earlier dominant, less self-confident and more cautious self-image of Korea as a nation that is still developing and that should therefore not risk economic growth. Conservative lawmaker Yu Gi-jun, for example, reminded his colleagues during the parliamentary debate about the "Framework Act on Low Carbon and Green Growth" that "[Korea] is a country that still needs growth [. . .]."[84] Proponents of this self-understanding called for a more cautious approach in contrast to a "Me First" leadership as envisioned by the Lee administration. For them, the international responsibility as a major carbon emitter was contrasted with the fact that Korea had no historical responsibility for high levels of greenhouse gases in the atmosphere.[85] However, as indicated already, this economic growth-focused, caution-advising strain of the "rising Korea" self-image lost traction domestically over the international success of the "green leadership" ambition.

Conclusion

The analysis of the wider political discourse of South Korea's international climate diplomacy showed the significant impact of President Lee's green leadership ambition through the international promotion of green growth. While the main role conception of being a "bridge" has been stable and dominant over the whole observation period, the inherent tension between Korea's global role model aspirations and its actual brown growth development track was only resolved through President Lee Myung-bak's focus on leading in international climate diplomacy through green growth promotion. Only then was Korea able to enact an internationally recognized role that fitted its own status and self-image. The aspect of the "rising Korea" self-image about being a country still in development and thus in need of prioritizing economic growth over climate change concerns made way for a green growth vision that still focused on economic opportunities but this time in low-carbon future-growth industries and markets. However, with a changing international context and a global momentum toward another climate accord, South Korea's hand was forced to secure its newfound status as a green leader in the post-Kyoto regime. The main goal now was to use the established green growth platforms and climate diplomatic channels to maintain and further expand the country's network and influence as a green leader and climate hub.

Figure 6.1 is an attempt to visualize these self-identifications as links between role conceptions and self-images. It is illustrative in that the initial vagueness of the "bridge" role conception relates to several self-images but only becomes manifested in a more meaningful role identity through the "green leadership" strategy of President Lee, building on the "rising Korea" self-image but ultimately transcending it.

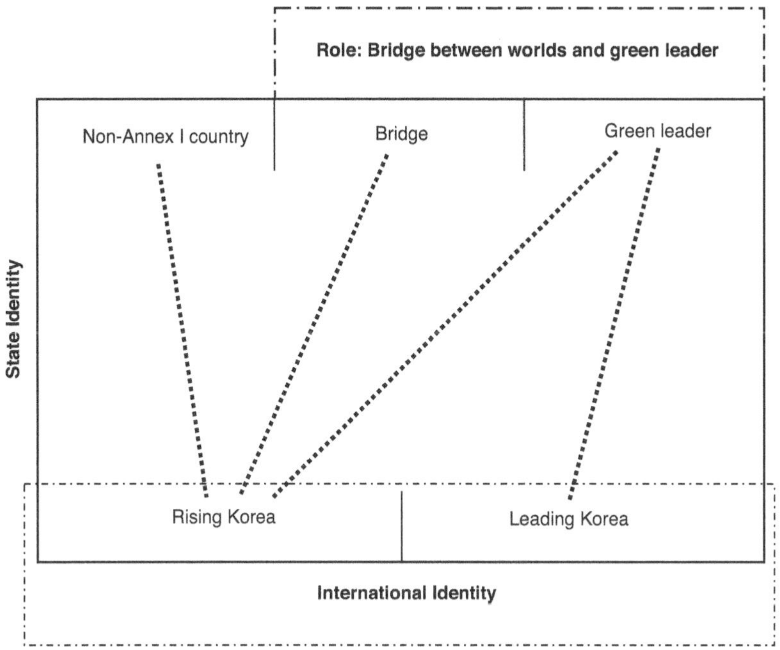

Figure 6.1 Roles, Role Conceptions and Self-images in ROK Climate Diplomacy, 2005–2016

Source: Created by author.

In the next chapter, these insights into the practices of self-identification prevalent in the case studies on international climate diplomacy and international peacekeeping are compared, analyzed, and put in context with the established literature on South Korea's identity narratives and foreign policy.

Notes

1. Saxer, Carl J. 2013: "Capabilities and Aspirations: South Korea's Rise as a Middle Power", *Asia Europe Journal*, Vol. 11, No. 4, pp. 397–413; Ikenberry, G. John and Mo, Jongryn 2013: *The Rise of Korean Leadership, Emerging Powers and International Order*, New York: Palgrave Macmillan.
2. *United Nations Framework Convention on Climate Change* (UNFCCC) 1992: *Treaty Text*, p. 4. Available at: https://unfccc.int/resource/docs/convkp/conveng.pdf [Accessed 5 November 2018].
3. Ibid.
4. Kim, Sungjin 2014: "South Korea's Climate Diplomacy: Analysis Based on the Perspective of 'Middle Power Diplomacy'", *EAI Middle Power Diplomacy Initiative Working Paper, No. 5*, Seoul: East Asia Institute, p. 1.
5. Yoon, Esook 2006: "South Korean Environmental Foreign Policy", *Asia-Pacific Review*, Vol. 13, No. 2, p. 75.

6. Watson, Iain 2012: "Contested Meanings of Environmentalism and National Security in Green Korea", *The Pacific Review*, Vol. 25, No. 5, p. 541.
7. Peritore, N. Patrick 1999: *Third World Environmentalism*, Case Studies from the Global South, Gainesville, FL: University Press of Florida, p. 91.
8. Yoon, E.S. 2006: p. 91.
9. Kim, S.J. 2014: p. 1.
10. Ikenberry, G.J. and Mo, J.R. 2013: p. 100.
11. Lee, Myung-bak 2012d: "Korea Has Emerged as a Nation That Takes the Initiative in Setting an Agenda Topic for Humanity, Which Is Well Demonstrated by the Selection of Korea as the Host Country of the GCF", *Speech*, 29 October.
12. Ikenberry, G.J. and Mo, J.R. 2013: p. 101.
13. *The World Bank 2012*: "Korea's Global Commitment to Green Growth", 3 May. Available at: www.worldbank.org/en/news/feature/2012/05/09/Korea-s-Global-Commitment-to-Green-Growth [Accessed 5 November 2018].
14. Organisation for Economic Co-Operation and Development (OECD) 2018: *Green Growth in Action: Korea*. Available at: www.oecd.org/korea/greengrowthinaction-korea.htm [Accessed 5 November 2018].
15. Blaxekjær, Lau 2016: "Korea as Green Middle Power: Green Growth Strategic Action in the Field of Global Environmental Governance", *International Relations of the Asia-Pacific*, Vol. 16, No. 3, pp. 443–476.
16. Ikenberry, G.J. and Mo, J.R. 2013: p. 110.
17. Kim, S.J. 2014; Kim, Inkyoung 2014: "Messages from a Middle Power: Participation by the Republic of Korea on Regional Environmental Cooperation on Transboundary Air Pollution Issues", *International Environmental Agreements: Politics, Law and Economics*, Vol. 14, No. 2, pp. 147–162; Han, Heejin 2015: "Korea's Pursuit of Low-Carbon Green Growth: A Middle-Power State's Dream of Becoming a Green Pioneer", *The Pacific Review*, Vol. 28, No. 5, pp. 731–754; Dent, Christopher M. 2015: "South Korea's Green Growth Strategy and East Asia's New Developmentalism", in: Anthony P. D'Costa (ed.), *After-Development Dynamics: South Korea's Contemporary Engagement with Asia*, Oxford: Oxford University Press, pp. 45–64; Watson, Iain 2015: "Middle Powers and Climate Change: The Role of KIA", *International Relations of the Asia-Pacific*, Vol. 15, No. 3, pp. 505–536.
18. Han, H.J. 2015.
19. Compare Watson, I. 2012; Yun, Sun-Jin, Ku, Dowan and Han, Jin-Yi 2013: "Climate Policy Networks in South Korea: Alliances and Conflicts", *Climate Policy*, Vol. 14, No. 2, pp. 283–301.
20. Dent, C.M. 2015.
21. Shim, David 2009: "'Green Growth': South Korea's Panacea?", in: Rüdiger Frank, Jim Hoare, Patrick Köllner and Susan Pares (eds.), *Korea 2010: Politics, Economy and Society*, Leiden: Brill, pp. 165–187; O'Donnell, Jill 2013: "Nuclear Power in South Korea's Green Strategy", *Green Growth Quarterly Update III-2013*, Council of Foreign Relations. Available at: www.files.ethz.ch/isn/166936/ROKGreenGrowthQuarterlyReport_June2013.pdf [Accessed 5 November 2018].
22. Han, H.J. 2015: p. 745.
23. Roh, Moo-hyun 2003: "A New Takeoff toward an Age of Peace and Prosperity", *Address by President Roh Moo-hyun at the 16th Inaugural Ceremony*, 25 February.
24. Song, Min-sun 2007a: "A Critical Turning Point of International Efforts against Climate Change", *The Korea Herald*, 21 September.
25. Ibid.
26. National Assembly Minutes 2009: *Special Committee on Climate Change Measures Session 18-282-5*, pp. 5–6.
27. Blaxekjær, L. 2016: p. 455.
28. Ibid.
29. Ibid.: p. 459.

30. Lee, M.B. 2012d.
31. Lee, Myung-bak 2010b: *The Keynote Speech to the Business for the Environment (B4E) Global Summit*, 22 April.
32. National Assembly Minutes 2009: *Special Committee on Climate Change Measures Session 18-282-5*, p. 26.
33. Ibid.: p. 18.
34. National Assembly Minutes 2009: *Special Committee on Climate Change Measures Session 18-282-6*, p. 24.
35. National Assembly Minutes 2009: *Special Committee on Climate Change Measures Session 18-282-5*, pp. 11–12.
36. Ibid.: p. 13.
37. Yoon, E.S. 2006: p. 75.
38. Lee, Myung-bak 2010a: *New Year Address to the Nation by the President*, 4 January.
39. Lee, Myung-bak 2011a: *Saving Energy in Everyday Life Represents an Effective Means for Overcoming High Oil Prices as Well as a Best Practice for Low-Carbon Green Growth*, 7 February.
40. Lee, Myung-bak 2008: "A Great People with New Dreams", *Address by President Lee Myung-bak on the 63rd Anniversary of National Liberation and the 60th Anniversary of the Founding of the Republic of Korea*, 15 August.
41. Ibid.
42. Lee, Myung-bak 2009c: *Address by President Lee Myung-bak on the 64th Anniversary of National Liberation*, 15 August.
43. Lee, Myung-bak 2009b: *Address to the Nation by President Lee Myungbak*, 13 July.
44. Lee, Myung-bak 2011c: *Address by President Lee Myung-bak on the 66th Anniversary of Liberation*, 15 August.
45. Lee, Myung-bak 2012b: *Address by President Lee on the 67th Anniversary of Liberation*, 15 August.
46. Lee, M.B. 2012d.
47. Ibid.
48. Ibid.
49. Lee, Myung-bak 2010d: "The G20 Seoul Summit Represents the Triumph of the Korean People and the Republic", *Radio Address*, 15 November.
50. Taylor, Lenore and Branigan, Tania 2014: "US and China Strike Deal on Carbon Cuts in Push for Global Climate Change Pact", *The Guardian*, 12 November. Available at: www.theguardian.com/environment/2014/nov/12/china-and-us-make-carbon-pledge [Accessed 5 November 2018].
51. National Assembly Minutes 2009: *Special Committee on Climate Change Measures Session 18-282-6*, p. 11.
52. National Assembly Minutes 2005: *Special Committee on Climate Change Session 17-256-4*, p. 12.
53. Ibid.: p. 11.
54. National Assembly Minutes 2009: *Special Committee on Climate Change Measures Session 18-282-6*, p. 11.
55. Ibid.: p. 12.
56. Ibid.: p. 1.
57. National Assembly Minutes 2009: *Special Committee on Climate Change Measures Session 18-282-5*, p. 35.
58. National Assembly Minutes 2009: *Plenary Session 18-285-1*, p. 36.
59. Ibid.
60. Lee, Myung-bak 2011d: "I Believe in the Potential of the Republic of Korea That Becomes Even Stronger in Times of Crisis", *Speech*, 3 October.
61. Park, Geun-hye 2013c: *Remarks by President Park Geun-hye at the Opening Ceremony of the Headquarters of the Green Climate Fund*, 4 December.

62. Park, Geun-hye 2014a: *Opening Remarks by President Park Geun-hye at the New Year Press Conference*, 6 January.
63. Park, Geun-hye 2014b: *Address by President Park Geun-hye at the 69th Session of the General Assembly of the United Nations*, 25 September.
64. National Assembly Minutes 2016: *Subcommittee for Judicial Review Session 20-346-2*, p. 5.
65. National Assembly Minutes 2016: *Plenary Session 20-346-11*.
66. National Assembly Minutes 2016: *Subcommittee for Judicial Review Session 20-346-2*, p. 17.
67. National Assembly Minutes 2016: *Standing Committee for Foreign Affairs Session 20-346-4*, p. 9.
68. National Assembly Minutes 2016: *Subcommittee for Judicial Review Session 20-346-2*, p. 17.
69. National Assembly Minutes 2016: *Plenary Session 20-346-11*, p. 4.
70. Ibid.: p. 5.
71. Lee, Myung-bak 2010c: "Marching Together toward a Greater Republic of Korea", *Address by President Lee Myung-bak on the 65th Anniversary of National Liberation*, 15 August.
72. Lee, M.B. 2012d.
73. Lee, M.B. 2010a.
74. Lee, M.B. 2008.
75. Ibid.
76. National Assembly Minutes 2009: *Special Committee on Climate Change Measures Session 18-282-5*, p. 26.
77. Lee, M.B. 2012d.
78. Young, Oran R. 2016: "IR in the Anthropocene", in: Kenneth Booth and Toni Erskine (eds.), *International Relations Theory Today*, Cambridge: Polity, p. 240.
79. Lee, M.B. 2009b.
80. Lee, M.B. 2010b.
81. Lee, Myung-bak 2011b: "We Need to Lay a Firm International Foundation for a Unified Korea, for an Advanced Green Growth Country", *Speech*, 16 May.
82. Park, Geun-hye 2016a: "Mutually Beneficial Partners for a New Future of Africa", *Address by H.E. Park Geun-hye President of the ROK at the African Union*, 27 May.
83. National Assembly Minutes 2012: *Standing Committee on Planning and Finance Session 19-311-5*, p. 7.
84. National Assembly Minutes 2009: *Special Committee on Climate Change Measures Session 18-282-5*, p. 44.
85. See also National Assembly Minutes 2009: *Special Committee on Climate Change Measures Session 18-282-6*, p. 24.

References

Blaxekjær, Lau 2016: "Korea as Green Middle Power: Green Growth Strategic Action in the Field of Global Environmental Governance", *International Relations of the Asia-Pacific*, Vol. 16, No. 3, pp. 443–476.

Dent, Christopher M. 2015: "South Korea's Green Growth Strategy and East Asia's New Developmentalism", in: Anthony P. D'Costa (ed.), *After-Development Dynamics: South Korea's Contemporary Engagement with Asia*, Oxford: Oxford University Press, pp. 45–64.

Han, Heejin 2015: "Korea's Pursuit of Low-Carbon Green Growth: A Middle-Power State's Dream of Becoming a Green Pioneer", *The Pacific Review*, Vol. 28, No. 5, pp. 731–754.

Ikenberry, G. John and Mo, Jongryn 2013: *The Rise of Korean Leadership, Emerging Powers and International Order*, New York: Palgrave Macmillan.

Kim, Inkyoung 2014: "Messages from a Middle Power: Participation by the Republic of Korea on Regional Environmental Cooperation on Transboundary Air Pollution Issues", *International Environmental Agreements: Politics, Law and Economics*, Vol. 14, No. 2, pp. 147–162.

Kim, Sungjin 2014: "South Korea's Climate Diplomacy: Analysis Based on the Perspective of 'Middle Power Diplomacy'", *EAI Middle Power Diplomacy Initiative Working Paper, No. 5*, Seoul: East Asia Institute.

Lee, Myung-bak 2008: "A Great People with New Dreams", *Address by President Lee Myung-bak on the 63rd Anniversary of National Liberation and the 60th Anniversary of the Founding of the Republic of Korea*, 15 August.

Lee, Myung-bak 2009a: *Address to the Nation by President Lee Myungbak*, 13 July.

Lee, Myung-bak 2009b: *Address by President Lee Myung-bak on the 64th Anniversary of National Liberation*, 15 August.

Lee, Myung-bak 2010a: *New Year Address to the Nation by the President*, 4 January.

Lee, Myung-bak 2010b: *The Keynote Speech to the Business for the Environment (B4E) Global Summit*, 22 April.

Lee, Myung-bak 2010c: "Marching Together toward a Greater Republic of Korea", *Address by President Lee Myung-bak on the 65th Anniversary of National Liberation*, 15 August.

Lee, Myung-bak 2010d: "The G20 Seoul Summit Represents the Triumph of the Korean People and the Republic", *Radio Address*, 15 November.

Lee, Myung-bak 2011a: *Saving Energy in Everyday Life Represents an Effective Means for Overcoming High Oil Prices as Well as a Best Practice for Low-Carbon Green Growth*, 7 February.

Lee, Myung-bak 2011b: "We Need to Lay a Firm International Foundation for a Unified Korea, for an Advanced Green Growth Country", *Speech*, 16 May.

Lee, Myung-bak 2011c: *Address by President Lee Myung-bak on the 66th Anniversary of Liberation*, 15 August.

Lee, Myung-bak 2011d: "I Believe in the Potential of the Republic of Korea That Becomes Even Stronger in Times of Crisis", *Speech*, 3 October.

Lee, Myung-bak 2012a: *Address by President Lee on the 67th Anniversary of Liberation*, 15 August.

Lee, Myung-bak 2012b: "Korea Has Emerged as a Nation That Takes the Initiative in Setting an Agenda Topic for Humanity, Which Is Well Demonstrated by the Selection of Korea as the Host Country of the GCF", *Speech*, 29 October.

National Assembly Minutes 2005: *Special Committee on Climate Change Session 17-256-4*.

National Assembly Minutes 2009a: *Plenary Session 18-285-1*.

National Assembly Minutes 2009b: *Special Committee on Climate Change Measures Session 18-282-5*.

National Assembly Minutes 2009c: *Special Committee on Climate Change Measures Session 18-282-6*.

National Assembly Minutes 2012: *Standing Committee on Planning and Finance Session 19-311-5*.

National Assembly Minutes 2016a: *Plenary Session 20-346-11*.

National Assembly Minutes 2016b: *Standing Committee for Foreign Affairs Session 20-346-4*.

National Assembly Minutes 2016c: *Subcommittee for Judicial Review Session 20-346-2*.

O'Donnell, Jill 2013: "Nuclear Power in South Korea's Green Strategy", *Green Growth Quarterly Update III-2013*, Council of Foreign Relations. Available at: www.files.ethz.ch/isn/166936/ROKGreenGrowthQuarterlyReport_June2013.pdf [Accessed 5 November 2018].

Organisation for Economic Co-Operation and Development (OECD) 2018: *Green Growth in Action: Korea*. Available at: www.oecd.org/korea/greengrowthinaction-korea.htm [Accessed 5 November 2018].

Park, Geun-hye 2013: *Remarks by President Park Geun-hye at the Opening Ceremony of the Headquarters of the Green Climate Fund*, 4 December.

Park, Geun-hye 2014a: *Opening Remarks by President Park Geun-hye at the New Year Press Conference*, 6 January.

Park, Geun-hye 2014b: *Address by President Park Geun-hye at the 69th Session of the General Assembly of the United Nations*, 25 September.

Park, Geun-hye 2016: "Mutually Beneficial Partners for a New Future of Africa", *Address by H.E. Park Geun-hye President of the ROK at the African Union*, 27 May.

Peritore, N. Patrick 1999: *Third World Environmentalism*, Case Studies from the Global South, Gainesville, FL: University Press of Florida.

Roh, Moo-hyun 2003: "A New Takeoff toward an Age of Peace and Prosperity", *Address by President Roh Moo-hyun at the 16th Inaugural Ceremony*, 25 February.

Saxer, Carl J. 2013: "Capabilities and Aspirations: South Korea's Rise as a Middle Power", *Asia Europe Journal*, Vol. 11, No. 4, pp. 397–413.

Shim, David 2009: "'Green Growth': South Korea's Panacea?", in: Rüdiger Frank, Jim Hoare, Patrick Köllner and Susan Pares (eds.), *Korea 2010: Politics, Economy and Society*, Leiden: Brill, pp. 165–187.

Song, Min-sun 2007a: "A Critical Turning Point of International Efforts against Climate Change", *The Korea Herald*, 21 September.

Taylor, Lenore and Branigan, Tania 2014: "US and China Strike Deal on Carbon Cuts in Push for Global Climate Change Pact", *The Guardian*, 12 November. Available at: www.theguardian.com/environment/2014/nov/12/china-and-us-make-carbon-pledge [Accessed 5 November 2018].

United Nations Framework Convention on Climate Change (UNFCCC) 1992: *Treaty Text*. Available at: https://unfccc.int/resource/docs/convkp/conveng.pdf [Accessed 5 November 2018].

Watson, Iain 2012: "Contested Meanings of Environmentalism and National Security in Green Korea", *The Pacific Review*, Vol. 25, No. 5, pp. 537–560.

Watson, Iain 2015: "Middle Powers and Climate Change: The Role of KIA", *International Relations of the Asia-Pacific*, Vol. 15, No. 3, pp. 505–536.

The World Bank 2012: "Korea's Global Commitment to Green Growth", 3 May. Available at: www.worldbank.org/en/news/feature/2012/05/09/Korea-s-Global-Commitment-to-Green-Growth [Accessed 5 November 2018].

Yoon, Esook 2006: "South Korean Environmental Foreign Policy", *Asia-Pacific Review*, Vol. 13, No. 2, pp. 74–96.

Young, Oran R. 2016: "IR in the Anthropocene", in: Kenneth Booth and Toni Erskine (eds.), *International Relations Theory Today*, Cambridge: Polity, pp. 231–249.

Yun, Sun-Jin, Ku, Dowan and Han, Jin-Yi 2013: "Climate Policy Networks in South Korea: Alliances and Conflicts", *Climate Policy*, Vol. 14, No. 2, pp. 283–301.

7 South Korea and Global Foreign Policy

Snapshots of Identity

The main question of this book is how South Korean international state identity relates to Seoul's global foreign policy in the 21st century, as there is, as yet, no comprehensive assessment of the international state identity of South Korea as an emerging global power in the academic literature. There are, however, various expectations for South Korea's future trajectory as an international actor, ranging from a future leader of the liberal international order to a building block for deeper regional integration and a resurgence of Korean nationalism. The three subquestions of this study are concerned with these potential identity narratives: Are there "globalist" self-understandings linked to roles like "global model" or "global citizen" that transcends national boundaries, embraces international law and limitations of national sovereignty, and adheres to international responsibilities as a "global citizen"? Or is the "globalist school" of Seoul's foreign policy simply about pursuing a narrower, more traditional "national interest" in the global arena? How about traditional postcolonial narratives that prioritize national sovereignty and "anti-imperialist" justice? And, last, did pan-Asianist narratives play any role in the legitimation of South Korea's global foreign policy?

In addition to this empirical gap, a conceptual gap has also been pointed out in that, amid various unconvincing conceptualizations of identity employed in the study of South Korean identity and foreign policy, where often identity was conflated with the concept role, most failed to account for the domestic Korean sense of self in the world. Hence, in this book there was a symbolic interactionist role-theoretical approach developed for the study of state identity in order to provide a conceptually sound and empirically rich agency-oriented account of Korea as an emerging global power. Through the focus on domestic legitimation and practices of self-identification underpinning the global foreign policy outreach, it was possible to retrieve self-images as pixelated snapshots of a broader international identity in relation to the respective role conceptions from case studies on two outstanding South Korean global foreign policy fields, international peacekeeping and climate diplomacy. So now, what are the identity narratives underpinning South Korean global foreign policy?

First of all, the study at hand suggests that there is no single identity in relation to Seoul's global foreign policy. By treating the state as a unitary actor and by categorizing actors simply according to their behavior, one not only misses the complexity of identity dynamics at play but also further obscures any comprehensive understanding of the state actor's present and future behavior to the outside world. It was helpful in this regard to introduce the concepts of role and role identities to the discussion of state identity and agency, and to think of state identity as two-layered with role conceptions linked to self-images. *Role* and *role identity* are terms that have often been conflated with *identity* in the literature. Whereas roles are the emergent relational outcomes of an ongoing process of interaction between the Self and the Other, the contestation over the "right" roles or role conceptions within the Self—the state as a composite actor—is about the legitimation of actions in reference to certain identity elements. As McCourt put it, "roles are the necessary social vehicle for action in its meaning-creating, identity-affirming sense."[1]

Hence, the self-identity of a state actor is not fixed in this understanding but is constantly formed through practices of self-identification or the linking of self-images and role conceptions by domestic political actors to make sense of and legitimate certain political actions. These self-images are employed to legitimate political action in domestic decision-making and are linked to specific role conceptions in the given social context. As the identity elements are generally undetermined, they can be linked to different role conceptions and to different outcomes, as was also apparent in the analysis at hand where, in the case study on climate diplomacy, the development-focused self-image "rising Korea" had been linked to the role conceptions "non–Annex I country," which advised against voluntary counter-measures to carbon emissions, and "bridge between worlds," where such voluntary measures were supported.

Layers, Narratives, and Self-Images

In the conceptualization of state identity developed here, there are two observable layers of the international self-identity of the state actor, ROK. These are the layer of concretely manifested role identities in a given social context and the more basic layer of self-images that have been employed for the legitimation of specific role conception and related actions. In the following section, this more comprehensive assessment of the state identity dynamics at play is presented and discussed and is put in context with the established literature on South Korean identity and foreign policy. The respective enacted roles and the related dominant role conceptions are treated as role identities, and the self-images utilized in the legitimation of specific foreign policy actions are treated as snapshots of the elusive corporate identity of South Korea as an international actor. The main self-images are that of a "rising Korea," focusing on Korea's elevated status as a result of economic development and democratization, on which the understanding of a "leading Korea" builds with a pronounced and confident leadership claim. Apart from these major self-images, there are slightly deviating and more marginal self-understandings, for example, where Korea's development is not yet successfully finished and still has to be safeguarded or the image of Korea as a peaceful nation that does not participate in the modern-day resource imperialism practiced by the US and China, as well as Japan.

As has been shown in the two case studies, there was a consensus over South Korea's appropriate role-playing in both case studies. The dominant role conception "responsible international member" was enacted as a stable role identity with support from two minor auxiliary role conceptions, "guardian of world peace" and "global model nation" in the case of Seoul's peacekeeping but with a rivaling role conception of "antimilitarist power." In the case of South Korean climate diplomacy, there was a rather vague and contested role conception of being a "bridge between worlds" over the whole observation period; however, this conception really only became meaningful and internationally recognized through its later transformation and manifestation in the "green leader" role conception during the presidency of Lee Myung-bak. Because of the dominance, stability (despite the transformation from "bridge" to "green leader"), and external recognition of the main role conceptions, it is plausible to attribute to South Korea the roles or role identities "responsible international member" in international peacekeeping, and "bridge between worlds" and "green leader" in climate diplomacy. These role identities could indeed be subsumed under a middle-power role set, but an investigation into international state identity should not stop here. In order to *understand* and not just *describe* South Korea as an international actor, it is necessary to focus on how these role identities have been legitimated with reference to self-images within the domestic decision-making process. Through the analysis of self-identification practices, it is possible to provide a comprehensive assessment not only of the dominant role conceptions and the related self-images but also of their contestation and the more marginal rivaling role conceptions and self-images that are also part of the actor's international state identity. In the following sections, these snapshots of the wider international identity are also presented and compared.

Rising Korea: Finally in the Global Spotlight

The main self-image over the whole observation period and linked to the dominant role conceptions "bridge between worlds" and "responsible member," as well as to the minor role conception "green responsible nation," is what has been termed the "rising Korea" self-image with narratives about Korea's hierarchical and geographical "in-betweenness" and its responsible advance from the periphery to the global center. Korea here is seen as a once peripheral and marginal country that is about to enter the ranks of advanced and relevant countries and has therefore found its "place in the sun" as a major player:

> The nation now stands side by side with other leading countries. Korea is no longer just a small speck on the global scene. It has become a major player. Korean-made cars and electronic goods are gaining accolades all

> around the world. [. . .] Korea has become a country that all nations around the world want to have closer ties with.[2]

The proud achievement of the successful development of Korea is at the core of this self-image, exemplified by the historical narratives about how Korea is the first country to transform from an aid recipient to a donor country, as well as "from beneficiary to benefactor."[3] In this self-image Korea is moving from the periphery to the center and now stands tall on the global stage, being proud of its achievements in economic development, democratization, and international diplomacy:

> Korea has transformed itself from an aid recipient to a donor despite all our difficulties. Ours is an unparalleled story in world history. Furthermore, the country has now become the chair of the G20 and will host the summit where leaders from around the world are to discuss the issues surrounding the economy and security. We are now squarely heading toward center stage from the periphery of the international arena.[4]

All three presidents likened Korea's and their region's rise to a journey from the periphery to the center stage of the global order. When President Roh Moo-hyun spoke of the past remoteness of Korea and Northeast Asia, he evoked a sense of regional identity and partnership with Asian neighbors in global affairs:

> As Northeast Asia is no longer on the periphery of the new international order, Korea is not a remote country any more. We will cast off our historic fate of being overshadowed and will instead come forward boldly and proudly to walk shoulder to shoulder with neighboring powers at the forefront of the new global order.[5]

President Lee, however, regularly referred to Korea's trajectory from the periphery to the center stage of the global order, without the regional dimension:

> Korea has taken on challenges with indomitable determination and worked hard to enter onto the center stage from the periphery of the international arena. [. . .] Over the past several years, Korea has taken a big step toward the center of the global stage. The country is now seizing the global spotlight like never before in its history.[6]

President Park Geun-hye, too, referred to this metaphor but related it to Northeast Asia's rise in world affairs as a whole:

> A mere few decades ago, East Asia was considered peripheral. Today, it is increasingly becoming the center of gravity of the world. This is clearly seen in the fact that the three East Asian countries—Korea,

China, and Japan—account for a quarter of global output and more than 20% of world trade.[7]

The country is aware of its risen status and knows about the responsibilities that come with it. In the case of peacekeeping, there was a strong notion about giving back to the UN and the international community after South Korea had been assisted and supported by UN troops during the Korean War. As Vice Foreign Minister Shin Kak-soo put it in 2010, "as a country that owes much to the UN, Korea has embarked on a path to contributing actively to the world community."[8] This narrative of historical debt played an important role in the case of legitimizing peacekeeping deployments and was regularly expressed by political actors throughout most parts of the political spectrum, apart from the political far-left.

Overall, there is a sense, however, of having been overlooked for too long, leading to an urge to utilize any opportunity to raise the visibility of Korea's rising status, for example, the hosting of international summits or conferences. The following quote from a speech by President Lee Myung-bak is a fitting illustration of these aspects:

> Over the past several years, Korea has taken a big step toward the center of the global stage. The country is now seizing the global spotlight like never before in its history. For the first time, Korea became a member of the G20, the premier economic forum in the world. It also became the first emerging and Asian country to host a G20 Summit in Seoul, which it did to great success.[9]

The global outreach and the taking-up of international commitments is hence a way of gaining self-assurance as a country of rising status at the center of the global order through the external recognition of the enactment of the related role identities "bridge between worlds" and "green leader" in climate diplomacy, as well as "responsible international member" in peacekeeping. In 2013 Foreign Minister Yun Byung-se even suggested that international journalists attending the World Journalist Conference in Seoul should relay the following images of Korea to the world:

> A "Secure Korea" undisturbed by the recent North Korean agitations, a "Credible Korea" that honors its promises, an "Empathetic Korea" that truly understands the hardships of others, a "Sharing and Caring Korea" that takes lead in development cooperation, and a "Responsible Korea" that contributes to regional and world peace. I am confident that at the end of your stay in Korea, you will be convinced of these attributes of Korea, as a safe, credible, empathetic, caring and responsible nation that is reaching out into the world. I hope you will join us as we strive to achieve these noble endeavors. As leaders that shape the global opinion, I hope you will widely tell the story of Korea's success and vision.[10]

Domestically, the external validation of the status aspirations and role identities has been regularly used to rally popular support for foreign policy positions and to legitimize the government's standing more generally.[11] The 2010 hosting of the G20 summit in Seoul is a case in point here. In the aftermath, President Lee called the summit a "triumph for of the Korean people and the Republic"[12] (and not just the city of Seoul) and applauded the bipartisan support in the National Assembly for the summit to "be turned into an epochal catalyst for the country's great advancement onto the center stage of the world."[13] Pride as an emotion was encouraged and individually expressed by decision-makers and international praise for Korea's performance to "global standards" was highlighted:

> The leadership and capabilities Korea has so far cultivated on the international scene were fully demonstrated on this occasion. Many talented people exhibited a high level of expertise that matches global standards. Many foreign professionals who worked together with Korean experts were deeply impressed and gave high marks to their abilities. Leaders of nations were also amazed at the seamless and convenient operation of meetings. [. . .] The Republic came to preside over a global summit and take the lead in reforming international financial organizations, including the IMF. We can now justifiably hold out our chests and step forward with self-confidence in the international arena.[14]

Last, it should be noted that in this self-image Korea's rise is still in the making, with the country still being seen as between truly advanced countries and the rest of the world. This is where the "dream of an advanced country" that has also been referred to in the literature on South Korea as a middle power[15] or the Global Korea project[16] connects the more widespread "rising Korea" self-image to the more confident and ambitious "leading Korea" self-image championed by President Lee Myung-bak's government, where the country has finally found entrance to the realm of advanced countries.

The significant Other for the "rising Korea" self-image is the "historical selves"[17] of the marginal and weak late Joseon as well as colonial Korea. It is here that the shrimp narrative about Korea's geopolitical vulnerability and marginalization is overcome not only by the projection of global relevance through contributions to international peace and security, as in the case of peacekeeping, but also in Korea's active agenda setting in climate diplomacy. As Foreign Minister Song Min-soon put it in his farewell speech in front of his ministry staff,

> [w]e must write the history of Korea by ourselves. The world history tells us that if a country cannot take care of its own matters, others will intrude. Korea is not a small country. Korea, as a middle power, can take a part in changing the world order related to our country by playing a due role. We have actually done that for the last few years, and the countries in our region as well as the international community have acknowledged our accomplishments.[18]

In this regard, Korea today is no longer the country that was turned away from the doors of the Second Hague Peace Conference a century earlier but is now the host of the biggest and most important international summits, such as the G20 or NSSs, themselves, with every international gathering, raising the visibility of Korea's advanced status to a global audience. Korea today literally cannot be overlooked anymore and its leaders have been consistently expressing this while conjuring up today's Korea's international agency:

> Korea today is not Korea of 100 years ago when it was divided into pieces sandwiched in the power struggles between China, Japan and Western powers. We now have sufficient power to maintain our history and territory. Depending on where the Korean people want to go, order in Northeast Asia is bound to be affected. The choice made by the Korean people will constitute to be essential in the business of predicting the future of Northeast Asia.[19]
>
> History is not pre-ordained. It is us who are making and creating it. Korea is no longer a shrimp surrounded by whales.[20]

Here, Korea is at the brink of advanced countries, the dream is about to become, true and Koreans see themselves with more confidence and more and more as masters of their own destiny:

> The kind of nation our patriotic forefathers intended to create was an upright, independent country, which carves out its own destiny. We have now been building up sufficient strength to be self-reliant and independent. Our economic power and defense prowess have come of age. We are no longer a peripheral country, which is one-sidedly swayed by developments in Northeast Asia.[21]

In a minor deviation of the rising Korea self-image, that was only referred to early in the observation period and during the initial debate surrounding President Lee's green leadership plans in 2008, Korea's development and international competitiveness were seen as still not strong enough to be risked over voluntary measures to limit carbon emissions. The narrative here was very much in line with what the literature on the matter had identified as the traditional prioritization of economic growth and argued that South Korea should not risk its international trading competitiveness vis-à-vis China and the US, who were also not restricted by the Kyoto Protocol regulations. Korea's lack of historical responsibility for the amount of carbon in the atmosphere as well as the country's corresponding official status as a non–Annex I (and therefore developing country) were raised as the main arguments.

Leading Korea: A Model for the World

The second major self-image present in both case studies is what I have called the "leading Korea" self-image, which was linked to the role conceptions "global model nation," "responsible member," "bridge," and "green leader." In many aspects, this self-image builds on the other major self-image "rising Korea" with a focus on Korea's successful development from being a small and weak country to a wealthy and influential nation in global affairs, but it goes one step further: here, Korea as a developed nation is either just at the brink of joining the exclusive club of "advanced countries" or has already made it. This self-image is closely related to the administration of President Lee Myung-bak and was, to a large part, the self-understanding linked to his Global Korea agenda. In this self-understanding, Korea offers itself as a model to other countries not just for its own economic development or domestic democratization but for its Me-first leadership with regard to pioneering a green growth model of development. It aims for the roles of mediator and agenda setter in order to establish Korea at the center of the global order, from where Korea with its "unique development experience and knowhow"[22] is poised to shape global affairs in the future, not as a follower but as a rule maker and "paragon for humankind."[23]

For this self-image, too, the significant Other is not so much a third country but a "historical self,"[24] this time the developmental South Korea of authoritarian president Park Chung Hee and the "miracle of the Han River." In contrast to the "rising Korea" self-image that emphasizes the international support for South Korea during and after the Korean War, under the self-image "leading Korea" there is a discursive shift to focus more on Korean achievements and the "unique experiences of overcoming colonial occupation, internecine war and poverty"[25] through Korean sacrifices and hard work. Therefore, it is unsurprising that President Park Geun-hye, his daughter, also tapped into the "beneficiary to benefactor" historical narrative to project a bridging and leading role internationally when she was linking the hard-earned developmental success to her father's decisions to send tens of thousands of Koreans to work as miners and nurses in Germany, as soldiers to Vietnam, or as construction workers and engineers to the Middle East from the 1960s to the 1970s:

> Some call [the South Korean economic development] the "Miracle on the Han River." But for those of us in Korea, it was anything but a miracle. And it wasn't just built from within. Koreans worked tirelessly in the mines of Germany, in the jungles of Vietnam and in the deserts of the Middle East.[26]
>
> "A smooth sea never made a skilled mariner." The words of this English proverb could not ring truer for Korea. For Korea has built up its strength from the crucible of the War and by weathering countless storms. Having once been on the receiving end of foreign aid, Korea is now a nation that gives. It is today a contributor both to fighting global poverty and to maintaining international peace.[27]

Furthermore, the international community represented anecdotally through tearful African leaders ("When I met them, they grabbed my hands and expressed their genuine thanks with tears filling their eyes.")[28] or international observers with only praise for Korea's contributions function as signifiers of external recognition of South Korea's international ambitions, which, in turn, is used for gaining domestic legitimacy for the very same measures:

> In this sense, Korea's unique experiences of overcoming colonial occupation, internecine war and poverty, can be used for the benefit of humanity. Dr. John Hamre of CSIS recently wrote about redefining Korea as a global leader. However, we in Korea should admit that there is a perception gap in how we view ourselves, and how the international community regards us in terms of our global role. As a recipient-turned-donor country, we need to contribute in a way that befits a country of our stature, and a divided nation with a thirst for greater peace as well as ultimate unification.[29]

Righteous Korea: A Peaceful Nation

Finally, the major deviant self-image is that of "righteous Korea," as observed in the peacekeeping case study, as the self-understanding put forward by the political left. In this self-image, Korea is an inherently peaceful nation that has achieved its heightened international status without doing harm to any other country. According to this self-image, "as a nation that has done no injury to another throughout history, Korea holds a sufficient moral ground and qualification to lead peace efforts in Northeast Asia."[30] This is at odds, however, with South Korea's participation in the Vietnam War, which jump-started the country's export-driven economic development under the military dictatorship of Park Chung-hee.[31] For the political left, the "peace-loving [Korean] people"[32] are misled into "resource imperialism" and military adventures through their affiliation, especially with the US,[33] potentially endangering the lives of Korean soldiers and ruining the good image of Korea in the Middle East, in particular.[34] The significant Other is correspondingly the US with its damaging influence on Korea and its supposed imperialist clout in the international community.[35] Hence, this self-image emphasizes more independence or autonomy in Korea's relationship with the US and beyond the international community.

No potential influence was observed of pan-Asianist identity elements on the global foreign policy in the fields of peacekeeping and climate diplomacy, which was one of the subquestions of this study. This might simply be because of the case selection, however. In the case study on peacekeeping, China has been framed as a "resource imperialist" power, just like the US or Japan by the far-left politicians.[36] Whereas the South Korean military proudly reported positive personal exchanges with Chinese soldiers in Lebanon through military diplomacy,[37] there is hardly a more complicated military cooperation partner imaginable for the South Korean military than Japan's Self Defense Forces, as illustrated by the political fallout of the "ammunition" incident in South Sudan.[38] In the case study on South Korean climate diplomacy, too, there were no substantial pan-Asian identity elements present. China, especially, was cast first as a major economic competitor and non–Annex I country under the Kyoto regime and then as a political rival with regard to green leadership after the US–China climate deal. It is possible, however, that a case study dealing with South Korean global developmental aid may yield different results regarding pan-Asian self-understandings.

Finally, the analysis turns to the related other two subquestions about the roles of traditional postcolonial principles that prioritize national sovereignty and anti-imperialism, as well as of a "globalist" self-understanding in the legitimation of South Korean global foreign policy in the 21st century. How has the traditionally strong Westphalian emphasis on securing sovereignty been linked to the global outreach, and did a post-Westphalian "globalist" self-understanding of sorts take hold in the global foreign policy discourse? As will be shown in the following section after a close reading of the historical narratives employed in these cases, the answer is—for now—that there has been no major self-understanding observed that somehow transcends national boundaries.

International Standing and National Autonomy

After a first glance at the retrieved practices of self-identifications, one notices the presence of seemingly normatively globalist, post-Westphalian role conceptions, such as "global model nation" and "responsible international member," and rather singular rhetoric references to global morality or international interdependence. However, a closer look at the related self-identification practices, the discursive linkages to the more basic level of self-images, leads to more traditional narratives of legitimation that prioritize sovereign autonomy and national advancement as key objectives of South Korea's global foreign policy. This argument rests on three aspects: the meaning of responsibility, the exhibited understanding of maturity, and the explicit objective to acquire international friends and knowledge against the backdrop of Korea's historical vulnerability.

First, when looking closer into the "responsible international member" role conception and the related self-images, it becomes apparent that these are dominated by decidedly national and not international narratives. In the "responsible international member" role conception and the related self-images "rising Korea" and "leading Korea," South Korean responsibility in the international community is discursively linked to external recognition or validation of Seoul's risen international status. On one hand, South Korea is acting out of a perceived duty to pay back a historical debt for the UN's support during and after the Korean War, but, on the other hand, Seoul sees itself as uniquely positioned to be a role model in the fields of development assistance and peacekeeping. Take this quote again from President Lee Myung-bak, for example:

> The Republic demonstrated that the assistance and collaboration from the international community was never wasted and that it was only right for

> such aid to be further expanded. In this regard, Korea was an exemplary beneficiary of international aid and cooperation in the 20th century. In the 21st century, the nation stands as a beacon for developing countries. Now is high time for the nation to return in full what it owes to the international community. Among other things, the Republic intends to share what it has experienced in achieving success with other countries across the globe.[39]

As a consequence, Korea's international responsibility is nothing more than Korea's national responsibility in the international realm. In the first case, the peacekeeping deployments have been rendered meaningful and appropriate as a grateful, national duty in order to repay historical debt, and, in the second case, the focus on national achievements in terms of development and defense makes peacekeeping a proud display of economic and military might, as well as of a certain generosity. Accordingly, in this framing from the "leading Korea" self-image, the country is not so much historically obliged to engage in UN peacekeeping but is uniquely qualified to share its own experience and viewpoints with the world, because Korea has something to offer to the world itself.

Second, the dream of Korea as an "advanced nation" (*seonjinguk*), as expressed and pursued by, for example, President Lee's Global Korea strategy, was aimed at making Korea a "mature global country" (*seongsukan segyegukga*). Even though mature behavior may be understood as "unselfish" international behavior that works toward the common good and not national interests, President Lee, for example, made clear that a mature foreign policy behavior is not a value in itself, but serves a further purpose: "A country can win global confidence only by becoming a mature member of the international community that shares and gives help to other nations going beyond establishing itself as an economic powerhouse."[40] This statement reduces maturity simply to a brand image and conducive to gaining "global confidence" for achieving a higher status and more influence.

Third, in addition to the widely referred to geopolitical dilemma Korea has found itself in as a shrimp, there was another argument raised in the wider political discourse about why Korea lost its sovereignty a century ago: a lack of international understanding and connections. Korean decision-makers explicitly referred to the late Joseon dynasty when presenting the lesson learned from history that—in order to secure autonomy and agency—a country has to reach out to the world. President Lee Myung-bak, for example, remembered Yu Kil-chun, the 19th-century intellectual and independence activist:

> In the late 19th century, Yu Kil-chun said that we should not be passive in opening up the country, but rather should take a more proactive role. Likewise, the 21st century calls on us to take on a more proactive role on the global stage.[41]

And at the 120th anniversary of the beginning of the Japanese-style *Gabo* reforms in 1894, Foreign Minister Yun Byung-se referred to the historical lesson that in order to be prepared against international power politics one must study international affairs and find allies:

> Sadly, Korea at the time did not possess the knowledge about the outside world nor the strength to protect itself. As a result, reform failed and led to the loss of its sovereignty. [. . .] Another lesson is the importance of diplomacy—making friends and neutralizing enemies. Throughout history, Korea was invaded by outside powers so many times. Throughout the 19th and 20th centuries, Korea learned the hard truth about the nature of power politics.[42]

This postcolonial perspective links back to the great importance that Korean leaders have attributed to the expansion and maintenance of international standing in both case studies on peacekeeping and climate diplomacy. Accordingly, the hosting of the GCF was to support because it was expected to "improve [. . .] the national brand image,"[43] and the extension of the mandate for deployed peacekeepers was important because it served the objective of raising international influence:

> Through the extensions, we hope to contribute not only to peace and stability in the region, but also to secure a strong base for diplomatic ties between those countries and Korea as well as to expand our influence in the international community through military contributions corresponding to our financial support.[44]

In this light, Korea's global outreach appears to be a way of securing the nation's sovereignty and agency through a global exercise of "know-your-enemy" and "make friends and neutralize enemies." Seen from this angle there is no post-Westphalian globalist self-understanding but a nationalist self-assertion from a postcolonial perspective. The same logic is evident, for example, in President Roh Moo-hyun's attempts to nurture a more independent (*jaju*) military stature vis-à-vis the US[45] and other regional powers, which is also part of the earlier discussed "righteous Korea" self-image on the political left:

> The kind of nation our patriotic forefathers intended to create was an upright, independent country, which carves out its own destiny. We have now been building up sufficient strength to be self-reliant and independent. Our economic power and defense prowess have come of age. We are no longer a peripheral country, which is one-sidedly swayed by developments in Northeast Asia. In light of our stature and role, it is reasonable for us to preserve our security with our own strength. We will do so without fail. Along with this effort, we will continue to nurture the ROK-U.S. alliance.[46]

Korea's global diplomatic outreach thus looks like a diversification of means to pursue the same interests of autonomy that has also been underpinning the ROK–US military alliance.

These findings are in line with what Park identified as "sovereign autonomy,"[47] the key source for political legitimacy as well a key feature in both South Korean and Japanese foreign and security policy. She argues that sovereign autonomy is "commonly understood as status to be negotiated through relations with other dominant powers in the international system. [. . . T]he constraints of hierarchy are a socially recognized fact and they provide meaning for leaders' actions and words."[48] According to her conceptual history of autonomy, political actors in both Japan and South Korea are contesting the meaning of the language of autonomy in their strategic discourses especially toward the US through framing vocabularies such as "independence," "self-reliance," or "autonomy," which are in effect indicating status-seeking through "insulating"[49] within the given international hierarchy. Then again, the different framing vocabulary of "advanced nations," "'international contributions' [. . .], global standards, national status, or position"[50] is pointing toward a more integrative-frame for a given international hierarchy. In addition to terms like *duty* and *responsibility*, these were exactly the terms at the core of the legitimation narratives supporting the role identity of "responsible member of the international community" (*gukjesahoeui irwon*) in the case study on peacekeeping. Still, Park overall concludes that

> the need to enhance status within an international hierarchy has surprisingly, and quite stubbornly, remained a resonant goal among Japanese and Koreans. This has remained the case despite structural transformations—including deimperialization, decolonization, rapid economic development, and democratization—that have pushed these two countries to become modern "Westphalian" states.[51]

The findings of the study at hand support Park's assessment for the case for South Korea's global foreign policy in the early 21st century, at least in the cases of peacekeeping and climate diplomacy. It is because of the absence of a pattern of self-identification that is transcending national boundaries with global commons–oriented or post-Westphalian globalist inclination and because of a strong focus on international standing and autonomy in the self-identification practices in the observed cases that lead to the conclusion that South Korean international state identity is dominated by the objective to maintain national autonomy above all in a changing regional and global context. This is not to say, however, that international identity changes toward more post-Westphalian globalist practices of self-identification, which would justify hopes for Seoul to be a liberal internationalist cornerstone, might not occur in the future.

On a different note, one might also wonder what this assessment of practices of self-identification and the importance of autonomy mean for a unified Korea in the future. Even though this remains for a large part conjecture, the example of reunified Germany provides some indications here, particularly with regard to overseas troop deployments such as peacekeeping dispatches. In a recently published study of public attitudes toward overseas troop deployments of the German military since reunification in 1990, Mader[52] concludes that there remained a significant, though weakening, difference between East and West Germans in terms of their support for the two salient norms of antimilitarism and Atlanticism in German foreign policy with regard to troop deployments. According to his findings,[53] for West Germans antimilitarism as a lesson learned from Nazi Germany's responsibility for the Second World War was largely compatible with the alignment with the West and NATO (*Westbindung*) and principles such as multilateralism or international responsibility (Atlanticism). In contrast, among East Germans the norm of antimilitarism against the backdrop of Nazi Germany's militarism was also supported strongly but the Atlanticist norm drastically less so. For an explanation, Mader points to the different socialization of East and West Germans and especially their different understandings of NATO as a mechanism of collective security on the one hand and an instrument of "imperialist" intervention on the other.[54]

So even in the (unlikely) case of a German-style reunification of the Korean Peninsula, where the South ROK peacefully absorbs the North state as an expanded democratic federation, the German example shows that a unified Korea will probably experience a growing role distance and changing practices of self-identification toward roles such as "responsible member of the international community" or "global model citizen." If we assume that in a unified Korea the "insulating"[55] status-seeking already present on the ROK's political left will be bolstered by isolationist and anti-imperialist attitudes held by a significant number of North Koreans especially toward Seoul's present-day allies, the US and Japan, overseas troop dispatches such as peacekeeping will be harder to legitimize with reference to alliance needs or international responsibilities. Of course, this extrapolation is little more than speculation; nevertheless, it may provoke more thinking about our assumptions about the future trajectory of a (reunified) Korea in the world.

Notes

1. McCourt, David M. 2012: "The Roles States Play: A Meadian Interactionist Approach", *Journal of International Relations and Development*, Vol. 15, No. 3, p. 370.
2. Lee, Myung-bak 2009c: *Address by President Lee Myung-bak on the 64th Anniversary of National Liberation*, 15 August.
3. Ha, Jong-Sik 2015: "ROK's Achievements through Its PKO Activities and Future Prospects", *Korean Institute for Defense Analysis, ROK Angle 122*, 29 May.
4. Lee, Myung-bak 2010c: "Marching Together toward a Greater Republic of Korea", *Address by President Lee Myung-bak on the 65th Anniversary of National Liberation*, 15 August.
5. Roh, Moo-hyun 2003: "A New Takeoff toward an Age of Peace and Prosperity", *Address by President Roh Moo-hyun at the 16th Inaugural Ceremony*, 25 February.
6. Lee, Myung-bak 2012b: *Address by President Lee on the 67th Anniversary of Liberation*, 15 August.
7. Park, Geun-hye 2014c: *Keynote Address at the 7th World Policy Conference*, 16 December.
8. Shin, Kak-soo 2010: *Keynote Address by H.E. Dr. Shin Kak-soo, Vice Minister of Foreign Affairs and Trade, at the Luncheon hosted by the UN Association of the Republic of Korea on UN Day*, 24 October.

9. Lee, M.B. 2012b.
10. Yun, Byung-se 2013: "Trustpolitik and Expanding Horizon of Korea's Foreign Policy", *Remarks by H.E. Yun Byung-se, Minister of Foreign Affairs, Republic of Korea, 2013 World Journalists Conference, Seoul*, 15 April.
11. Compare Von Soest, Christian and Grauvogel, Julia 2017: "Identity, Procedures and Performance: How Authoritarian Regimes Legitimize Their Rule", *Contemporary Politics*, Vol. 23, No. 3, pp. 287–305.
12. Lee, Myung-bak 2010d: "The G20 Seoul Summit Represents the Triumph of the Korean People and the Republic", *Radio Address*, 15 November.
13. Ibid.
14. Ibid.
15. Saxer, Carl J. 2013: "Capabilities and Aspirations: South Korea's Rise as a Middle Power", *Asia Europe Journal*, Vol. 11, No. 4, pp. 397–413.
16. John, Jojin V. 2015: "Globalization, National Identity and Foreign Policy: Understanding 'Global Korea'", *The Copenhagen Journal of Asian Studies*, Vol. 33, No. 2, pp. 38–57.
17. Harnisch, Sebastian 2016b: "China's Historical Self and Its International Role", in: Sebastian Harnisch, Sebastian Bersick and Jörn-Carsten Gottwald (eds.), *China's International Roles*, London/New York: Routledge, pp. 38–58.
18. Song, Min-sun 2008: *Farewell Speech by H.E. Song Min-soon Minister of Foreign Affairs and Trade of the Republic of Korea*, 29 February.
19. Roh, Moo-hyun 2004b: *Address by President Roh Moo-hyun on the 59th Anniversary of National Liberation*, 15 August.
20. Yun, Byung-se 2014: "The Future of History: A New Korea, a New Asia, a New World" *Keynote Speech by H.E. Yun Byung-se Minister of Foreign Affairs at the Asan Plenum 2014*, 22 April.
21. Roh, Moo-hyun 2004a: *Remarks by President Roh Moo-hyun on the 49th Memorial Day*, 6 June.
22. Lee, Myung-bak 2012c: *President Lee Myung-bak's Address to the National Assembly Proposing the Government's Budget Plan for FY 2013 and Plans for Fiscal Operations*, 5 October.
23. Lee, Myung-bak 2009a: *New Year Policy Address to the Nation by President Lee Myung-bak*, 2 January.
24. Harnisch, S. 2016b.
25. Yun, Byung-se 2015: *Remarks by H.E. Yun Byung-se, Minister of Foreign Affairs, at UN Day Reception*, 23 October.
26. Park, Geun-hye 2013a: *Address by President Park Geun-Hye of the Republic of Korea to the Joint Session of the United States Congress*, 9 May.
27. Park, Geun-hye 2013b: *Remarks by President Park Geun-hye at the Parliament of the United Kingdom*, 11 May.
28. Lee, M.B. 2010d.
29. Yun, B.S. 2015.
30. Roh, Moo-hyun 2007: *Address by President Roh Moo-hyun on the 88th March First Independence*, 1 March.
31. Breuker, Remco E. 2009: "Korea's Forgotten War: Appropriating and Subverting the Vietnam War in Korean Popular Imaginings", *Korean Histories*, Vol. 1, No. 1, pp. 40–41.
32. National Assembly Minutes 2009: *Plenary Session 18-285-1*, p. 18.
33. National Assembly Minutes 2012: *Standing Committee on Foreign Affairs Session 19-311-9*, p. 22.
34. National Assembly Minutes 2009: *Plenary Session 18-283-1*, p. 8.
35. National Assembly Minutes 2006: *Plenary Session 17-263-1*, p. 51.
36. Compare National Assembly Minutes 2012: *Standing Committee on Foreign Affairs Session 19-311-9*, p. 22.

37. National Assembly Minutes 2008: *Special Committee on Reviewing the Extension of the UNIFIL Mandate Session 18-276-1*, p. 7.
38. Fischer, Max 2013: "Japan and South Korea Can't Even Cooperate over Peacekeeping in South Sudan", *The Washington Post*, 26 December. Available at: www.washingtonpost.com/news/worldviews/wp/2013/12/26/japan-and-south-korea-cant-even-cooperate-over-peacekeeping-in-south-sudan/?utm_term=.edddfccef050 [Accessed 5 November 2018]; Ha, Eo-young and Seok, Jin-hwan 2013: "S. Korean Troops Borrow Japanese Ammunition in South Sudan", *The Hankyoreh*, 31 December. Available at: http://english.hani.co.kr/arti/english_edition/e_international/617788.html [Accessed 5 November 2018].
39. Lee, Myung-bak 2009d: *Radio and Internet Address to the Nation by President Lee Myung-bak*, 30 November.
40. Lee, Myung-bak 2013: *The Nation Brand Value Soaring: From Korea Discount to Korea Premium*, 21 January.
41. Lee, M.B. 2009c.
42. Yun, B.S. 2014.
43. National Assembly Minutes 2012: *Standing Committee on Planning and Finance Session 19-311-5*, p. 7.
44. National Assembly Minutes 2011: *Standing Committee on Foreign Affairs Session 18-304-2*, p. 4.
45. Ryu, Yongwook 2013: "Does Political Generation Matter for Foreign Policy? South Korea's Foreign Policy Changes, 2003 to 2007", *The Korean Journal of International Studies*, Vol. 11, No. 1, pp. 117–118.
46. Roh, M.H. 2004a.
47. Park, Seo-Hyun 2017: *Sovereignty and Status in East Asian International Relations*, Cambridge: Cambridge University Press.
48. Ibid.: p. 22.
49. Ibid.: p. 103.
50. Park, Seo-Hyun 2013: "Changing Definitions of Sovereignty in Nineteenth-Century East Asia: Japan and Korea between China and the West", *Journal of East Asian Studies*, Vol. 13, No. 2, p. 17.
51. Park, S.H. 2017: pp. 12–13.
52. Mader, Matthias 2017: *Öffentliche Meinung zu Auslandseinsätzen der Bundeswehr: Zwischen Antimilitarismus und transatlantischer Orientierung [Public Opinion towards Overseas Troop Deployments of the German Military: Between Anti-Militarism and Transatlantic Orientation]*, Wiesbaden: Springer VS.
53. Ibid.: p. 196.
54. Ibid.: p. 90.
55. Park, S.H. 2017: p. 103.

References

Breuker, Remco E. 2009: "Korea's Forgotten War: Appropriating and Subverting the Vietnam War in Korean Popular Imaginings", *Korean Histories*, Vol. 1, No. 1, pp. 36–59.

Fischer, Max 2013: "Japan and South Korea Can't Even Cooperate over Peacekeeping in South Sudan", *The Washington Post*, 26 December. Available at: www.washingtonpost.com/news/worldviews/wp/2013/12/26/japan-and-south-korea-cant-even-cooperate-over-peacekeeping-in-south-sudan/?utm_term=.edddfccef050 [Accessed 5 November 2018].

Ha, Eo-young and Seok, Jin-hwan 2013: "S. Korean Troops Borrow Japanese Ammunition in South Sudan", *The Hankyoreh*, 31 December. Available at: http://english.hani.co.kr/arti/english_edition/e_international/617788.html [Accessed 5 November 2018].

Ha, Jong-Sik 2015: "ROK's Achievements through Its PKO Activities and Future Prospects", *Korean Institute for Defense Analysis, ROK Angle 122*, 29 May.

Harnisch, Sebastian 2016b: "China's Historical Self and Its International Role", in: Sebastian Harnisch, Sebastian Bersick and Jörn-Carsten Gottwald (eds.), *China's International Roles*, London/New York: Routledge, pp. 38–58.

John, Jojin V. 2015: "Globalization, National Identity and Foreign Policy: Understanding 'Global Korea'", *The Copenhagen Journal of Asian Studies*, Vol. 33, No. 2, pp. 38–57.

Lee, Myung-bak 2009a: *New Year Policy Address to the Nation by President Lee Myung-bak*, 2 January.

Lee, Myung-bak 2009b: *Address by President Lee Myung-bak on the 64th Anniversary of National Liberation*, 15 August.

Lee, Myung-bak 2009c: *Radio and Internet Address to the Nation by President Lee Myung-bak*, 30 November.

Lee, Myung-bak 2010a: "Marching Together toward a Greater Republic of Korea", *Address by President Lee Myung-bak on the 65th Anniversary of National Liberation*, 15 August.

Lee, Myung-bak 2010b: "The G20 Seoul Summit Represents the Triumph of the Korean People and the Republic", *Radio Address*, 15 November.

Lee, Myung-bak 2012a: *Address by President Lee on the 67th Anniversary of Liberation*, 15 August.

Lee, Myung-bak 2012b: *President Lee Myung-bak's Address to the National Assembly Proposing the Government's Budget Plan for FY 2013 and Plans for Fiscal Operations*, 5 October.

Lee, Myung-bak 2013: *The Nation Brand Value Soaring: From Korea Discount to Korea Premium*, 21 January.

Mader, Matthias 2017: *Öffentliche Meinung zu Auslandseinsätzen der Bundeswehr: Zwischen Antimilitarismus und transatlantischer Orientierung [Public Opinion towards Overseas Troop Deployments of the German Military: Between Anti-Militarism and Transatlantic Orientation]*, Wiesbaden: Springer VS.

McCourt, David M. 2012: "The Roles States Play: A Meadian Interactionist Approach", *Journal of International Relations and Development*, Vol. 15, No. 3, pp. 370–392.

National Assembly Minutes 2006: *Plenary Session 17-263-1*.

National Assembly Minutes 2008: *Special Committee on Reviewing the Extension of the UNIFIL Mandate Session 18-276-1*.

National Assembly Minutes 2009a: *Plenary Session 18-283-1*.

National Assembly Minutes 2009b: *Plenary Session 18-285-1*.

National Assembly Minutes 2011: *Standing Committee on Foreign Affairs Session 18-304-2*.

National Assembly Minutes 2012a: *Standing Committee on Foreign Affairs Session 19-311-9*.

National Assembly Minutes 2012b: *Standing Committee on Planning and Finance Session 19-311-5*.

Park, Geun-hye 2013a: *Address by President Park Geun-Hye of the Republic of Korea to the Joint Session of the United States Congress*, 9 May.

Park, Geun-hye 2013b: *Remarks by President Park Geun-hye at the Parliament of the United Kingdom*, 11 May.

Park, Geun-hye 2014: *Keynote Address at the 7th World Policy Conference*, 16 December.

Park, Seo-Hyun 2013: "Changing Definitions of Sovereignty in Nineteenth-Century East Asia: Japan and Korea between China and the West", *Journal of East Asian Studies*, Vol. 13, No. 2, pp. 281–307.

Park, Seo-Hyun 2017: *Sovereignty and Status in East Asian International Relations*, Cambridge: Cambridge University Press.

Roh, Moo-hyun 2003: “A New Takeoff toward an Age of Peace and Prosperity”, *Address by President Roh Moo-hyun at the 16th Inaugural Ceremony*, 25 February.

Roh, Moo-hyun 2004a: *Remarks by President Roh Moo-hyun on the 49th Memorial Day*, 6 June.

Roh, Moo-hyun 2004b: *Address by President Roh Moo-hyun on the 59th Anniversary of National Liberation*, 15 August.

Roh, Moo-hyun 2007: *Address by President Roh Moo-hyun on the 88th March First Independence*, 1 March.

Ryu, Yongwook 2013: “Does Political Generation Matter for Foreign Policy? South Korea’s Foreign Policy Changes, 2003 to 2007”, *The Korean Journal of International Studies*, Vol. 11, No. 1, pp. 113–141.

Saxer, Carl J. 2013: “Capabilities and Aspirations: South Korea’s Rise as a Middle Power”, *Asia Europe Journal*, Vol. 11, No. 4, pp. 397–413.

Shin, Kak-soo 2010: *Keynote Address by H.E. Dr. Shin Kak-soo, Vice Minister of Foreign Affairs and Trade, at the Luncheon Hosted by the UN Association of the Republic of Korea on UN Day*, 24 October.

Song, Min-sun 2008: *Farewell Speech by H.E. Song Min-soon Minister of Foreign Affairs and Trade of the Republic of Korea*, 29 February.

Von Soest, Christian and Grauvogel, Julia 2017: “Identity, Procedures and Performance: How Authoritarian Regimes Legitimize Their Rule”, *Contemporary Politics*, Vol. 23, No. 3, pp. 287–305.

Yun, Byung-se 2013: “Trustpolitik and Expanding Horizon of Korea’s Foreign Policy”, *Remarks by H.E. Yun Byung-se, Minister of Foreign Affairs, Republic of Korea, 2013 World Journalists Conference, Seoul*, 15 April.

Yun, Byung-se 2014: “The Future of History: A New Korea, a New Asia, a New World”, *Keynote Speech by H.E. Yun Byung-se Minister of Foreign Affairs at the Asan Plenum 2014*, 22 April.

Yun, Byung-se 2015: *Remarks by H.E. Yun Byung-se, Minister of Foreign Affairs, at UN Day Reception*, 23 October.

8 Conclusion

Dream of Autonomy

This book set out to explore the identity in "transformation"[1] or "at crossroads"[2] of the ROK as an emerging global player, as there is no comprehensive study yet about how South Korea as a country relates itself to a globalized world beyond the usually covered bilateral relationships with, for example, North Korea, China, Japan, and the US. The main research interest was about the relationship between the international state identity of South Korea of the early 21st century and Seoul's global foreign policy in the areas of peacekeeping and climate diplomacy. In order to undertake such an investigation into international state identity in relation to foreign policy, this book presents a symbolic interactionist role-theoretical conceptualization of international state identity understood as a set of discursive practices of self-identification for the domestic legitimation of foreign policy actions. The state is understood as a composite actor here, a social structure within which domestic political actors contest over the "right" course for their polity by employing different role conceptions linked to self-images in order to legitimize political actions as purposeful actions. This study offers a novel empirical approach to the study of state identity in IR as a contribution to the theoretical development of identity and its relationship to role in IR and FPA.

Besides the conceptual-theoretical contribution, the study at hand provides a comprehensive and agency-oriented account of the relationship between international state identity and global foreign policy of South Korea. With regard to the main research question of how South Korean international state identity relates to Seoul's global foreign policy in the 21st century, this study has shown that there is no single unitary state identity underpinning South Korea's global outreach. Korea was able to successfully enact distinct roles as a "responsible international member" in peacekeeping and as "bridge between worlds" through the "green leader" role identity in climate diplomacy, but these are case-dependent and only the outer layer of South Korean international state identity. The inner layer consists of self-images on which different role conceptions and eventually established role identities are built. These self-images that are utilized in the legitimation of specific foreign policy actions are treated as snapshots of the elusive broader international identity of South Korea. One of the key findings of this study is accordingly the account of the three major self-images that have been termed "rising Korea," "leading Korea," and "righteous Korea."

In the most dominant self-image "rising Korea," the country is moving from the periphery to the center and now stands tall on the global stage, being proud of its achievements in economic development, democratization, and international diplomacy. The global outreach is here a way of gaining self-assurance as a country of rising status at the center of the global order through heightened visibility and external recognition, against the background of Korea's own colonial experience, geopolitical vulnerability, and marginalization. The self-image of a "leading Korea" builds on the previous image with a more pronounced and confident international leadership claim. Here, Korea offers itself as a model to other countries for its own economic development, domestic democratization, and leadership, especially with regard to pioneering a green growth model of development. It aims to firmly establish Korea at the center of the global order, from where Korea is poised to shape global affairs in the future, not as a follower but as a rule maker. Global engagement is here a way of raising the country's international status and brand even further and to project Korean norms and values in the international arena. Finally, there was also a minor, deviant self-image put forward by the political left, for which South Korea's morally and peacefully "pure" state is tainted by the association with the US and unjust international power politics.

In contrast to other studies that are either simply assuming a unitary actor, focusing solely on government views or conflating role and identity, the approach developed here also accounts for rivaling role conceptions and deviant self-images. In the case study on peacekeeping, for example, it was shown that there is a distinctly different role conception of "antimilitarist power" and a related self-image about a "righteous Korea," in which inherent peacefulness is betrayed by military overseas deployments in support of the US-led international order. The research has also shown that at the core of South Korea's relationship with global affairs lies the objective of expanding Seoul's international standing in order to assure national autonomy. Global foreign policy has been a means of validating the risen international status of the country and a way of trying to establish South Korea as a rule maker rather than a rule follower. Its legitimation was dominated by narratives about fostering a premium nation brand and extensive ties around the world, against a referenced historiography of vulnerability due to the lack of international connections and knowledge. The most recent global outreach, whether pursued by conservative or progressive administrations, of the emerging power Korea is thus not simply exemplary middle powermanship[3] in support of the existing liberal international order[4] but more a way of integrating into a given international hierarchy in order to secure what Park called "sovereign autonomy."[5] This follows from an intertextual and cross-sectional reading of the retrieved self-images of the political debate in the observed cases and does not mean that there are no political actors currently putting forward a self-image of Korea as a "global citizen," where the self-understanding is transcending national borders. There may well be actors in civil society, the media, the bureaucracy, or the political parties, who support such a post-Westphalian globalist self-image, in general, but it has yet to be activated politically with regard to legitimizing foreign policy action in the two observed policy fields. This limitation due to scope and case selection is the main limitation of this study. The self-images as snapshots of the broader South Korean international identity are, as a consequence, rather "coarsely pixelated" but insightful nonetheless.

For a more zoomed-in or panoramic account of South Korean international identity, in general, future research might investigate other policy fields as cases, such as overseas developmental assistance or global trade policy, but also different levels of analysis as, for example, civil society or media discourses, either separately or in addition to the political discourse. Ultimately it could also be worthwhile to turn to popular culture discourses and explore how self-identification occurs in novels, comics, movies, and TV dramas, but—for the role-theoretical approach at hand—this would be relevant politically only indirectly. Following the "aesthetic turn"[6] in IR, popular texts are not just a mere reflection of a political reality but are also shaping social reality, meanings, values, identities, and thus politics. Apart from cultural representations manufactured explicitly for political purposes, Grayson, Davis, and Philpott point out that "[b]eyond formal policy statements, press interviews and traditional forms of propaganda, popular culture draws attention to how understandings of world politics and the legitimation of policy postures can be generated."[7] Consequently, politicians have to draw from cultural resources to make sense of political options and decisions not only for themselves but also for their constituencies.[8] It is the main focus of these approaches to examine the workings of power involved in the production of meaning through cultural artifacts, in which a cultural text can both support and/or subvert the established power relations.[9] For example, when former South Korean president Park Geun-hye was meeting South Korean peacekeepers from the Hanbit Unit in South Sudan in May 2016, she praised them as the "real descendants of the sun" in front of the TV cameras,[10] referring to the very popular and successful TV drama of the same name. *Descendants of the Sun* aired in early 2016 and stands out as one of the first mainstream popular cultural imaginations of South Korea as a developed country and peacekeeping actor in a distant country. In the drama's representation, the ROK is a rich and good member of the international community, naturally contributing to the security and development of the fictional country Urk, somewhere in the Balkans, through the deployment of peacekeepers and medical volunteers and the construction of a solar power plant. But through reading the drama in the context of "real-world" South Korean global politics and focusing on which aspects of reality might have been normalized or reinforced, one can gain further insights into the popular cultural underpinnings of current South Korean state identity and global foreign policy that political leaders tap into.[11]

For example, the usually-referred-to historical context of the UN support for South Korea during and after the Korean War and the related narrative of "paying back" the international community as a "child of the United Nations" is noticeably absent in *Descendants of the Sun*. Furthermore, throughout the series, the UN is only represented through some flags in the background and without the typical peacekeeping signifiers such as the iconic UN blue helmets or scarves missing from the South Korean and any other military uniforms. In contrast, the presence of South Korean military troops in the fictional country is never challenged by the drama's characters but is instead experienced as a "normal" reflection of the country's risen status. Characters explicitly voice their positive emotions and patriotism in view of Korean-language signage in distant lands, whether on a Korean-built futuristic solar power plant or on donated T-shirts worn by poor local children. In the context of the "leading Korea" self-image that was retrieved in the study at hand, the South Korea of *Descendants of the Sun* appears to be represented as a shiny premium brand and "paragon for humankind,"[12] acting not to repay a historical debt but to display a national achievement untainted by Korea's own dark chapters of history.

Another further research area is the dimension of emotion in the self-identification practices of South Korean state identity and global foreign policy. Because of the still largely theoretical debate about emotions in world politics and the corresponding lack of empirically applicable conceptualizations of emotions and state identities, this identity dimension was not part of the approach presented here. As illustrated in the case study on South Korean peacekeeping, however, there have been discursive features such as emotion terms, connotations, analogies, and metaphors[13] such as pride, when President Lee Myung-bak praised the "global standard"[14] of South Korea's (and not Seoul's) hosting of the G20 summit in 2010, and embarrassment, when Representative Hwang Jin-ha scolded the government for a lack of ambition to put more South Korean officers in leadership roles of international peacekeeping missions.[15] Following Koschut, these emotional features may be worth further exploration with regard to the discursive practices of emotional Othering as well as the performativity and interpellation of emotions.[16] The research task here would be to move analytic focus "from tracing the meaning of single or multiple emotion words to contextualizing their meaning by looking at how these expressions are directed at and resonate with particular audiences."[17] Another way of approaching the emotional dimension of state identity is to focus on group-level emotions, "where one feels guilty not for something one did oneself but for something one's group did."[18] Mercer, for example, argues that identity always depends on social emotions, because "[i]dentification requires a feeling of attachment; it is intrinsically social."[19]

In terms of policy relevance, this study teaches us caution with regard to drawing conclusions from assumed politico-cultural identities or cultural norms of countries, as these are neither zombie-like executioners of alleged national characters nor dull norm dupes complying with norms once they have been internalized. As the symbolic role-theoretical approach here suggests, a state actor's self-identification is a constant performative and narrative (but also emotional) process in which political decision-makers have sometimes more agency, sometimes less. Therefore, if one is interested in either the foreign policy behavior of any given state or the stability of the order emerging from different role-playing actors, a look at the domestic self-identification practices is advisable, as illustrated with this study at hand.

With regard to the new administration of President Moon Jae-in, who was inaugurated in May 2017 after the impeachment of President Park Geun-hye, it is too early to draw any conclusions as to which global role identities his government will pursue and try to enact. If his early speeches at the G20 summit in Hamburg in July 2017 are any indication, he follows his predecessors in seeing climate change response as "an investment in future generations and also an opportunity to create new jobs and growth engines."[20] He also set a new tone with regard to the international refugee crisis by referring to Korean history and the many refugees from the Korean War, which is also part of his personal family history. His parents were among the 14,000 North Korean refugees evacuated on board the US cargo freighter *SS Meredith Victory* from the besieged port of Hungnam in December 1950.[21] Moreover, his administration initially seemed strongly focused on domestic issues, such as institutional reforms and job creation, announcing[22] the phasing-out of the country's nuclear reactors because of safety concerns in the context of the Fukushima nuclear accident in 2011 and the strongest ever earthquake in the history of Korea in late 2016, despite South Korea's strong reliance on this energy resource and the internationally very competitive nuclear export industry. Since the tensions of the summer of 2017 between US president Trump and North Korean leader Kim Jong-un, however, and Pyongyang's successful testing of the Hwasong-14 intercontinental ballistic missile, President Moon very much focused his foreign policy efforts during his first year in office on North Korean engagement and summit diplomacy with Washington and Pyongyang.

Furthermore, this study bridges a gap between Korean Studies and IR, as it is dedicated to the question of how Korea confronts globalization on the identity and cultural level but with a focus on state and political identities and with IR theory and methodology applied. Besides, for the discipline of IR, this study can serve as an example of how to think about South Korea as a foreign policy actor by taking Korean agency and domestic self-identification practices more seriously and less so outsider's normative strategic biases. The agency-oriented perspective at hand treats identity as a fragile and continuously contested "sense of self" that has to be constantly manifested through role enactment and promises to complement more systemic identity approaches, such as the literature on norms and socialization. Finally, for the discipline of Korean Studies, this study has illustrated that IR approaches can also contribute to our understanding of Korean collective identity and globalization, in addition to the political-strategic questions about, for example, inter-Korean relations or the impact of the different missile defense systems on a regional "balance of power" that are prevalent in foreign policy–focused studies on Korea.

Notes

1. Ikenberry, G. John and Mo, Jongryn 2013: *The Rise of Korean Leadership, Emerging Powers and International Order*, New York: Palgrave Macmillan, pp. 164–165; Denney, Steven and Friedhoff, Karl 2013: "South Korea and a New Nationalism in an Era of Strength and Prosperity," *Center for Strategic and International Studies PacNet Newsletter 75*, 7 October. Available at: https://csis-prod.s3.amazonaws.com/s3fs-public/legacy_files/files/publication/Pac1375_0.pdf [Accessed 5 November 2018].
2. Rozman, Gilbert 2012: "South Korean National Identity: A Six-Dimensional Analysis", in: Gilbert Rozman (ed.), *East Asian National Identities: Common Roots and Chinese Exceptionalism*, Stanford: Stanford University Press, p. 67; Snyder, Scott A. 2018: *South Korea at the Crossroads: Autonomy and Alliance in an Era of Rival Powers*, New York: Columbia University Press.
3. Cooper, Andrew F. 2015: "Soft Power and the Recalibration of Middle Powers: South Korea as an East Asian Leader and Canada as the Exemplar of the Traditional Model", in: Jan Melissen and Yul Sohn (eds.), *Understanding Public Diplomacy in East Asia: Middle Powers in a Troubled Region*, New York: Palgrave Macmillan, p. 32.
4. Ikenberry, G.J. and Mo, J.R. 2013: pp. 164–165.
5. Park, Seo-Hyun 2013: "Changing Definitions of Sovereignty in Nineteenth-Century East Asia: Japan and Korea between China and the West", *Journal of East Asian Studies*, Vol. 13, No. 2, pp. 281–307.
6. Bleiker, Roland 2001: "The Aesthetic Turn in International Political Theory", *Millenium: Journal of International Studies*, Vol. 30, No. 3, pp. 509–533.
7. Grayson, Kyle, Davies, Matt and Philpott, Simon 2009: "Pop Goes IR? Researching the Popular Culture-World Politics Continuum", *Politics*, Vol. 29, No. 3, p. 159.
8. Weldes, Jutta 2003: "Popular Culture, Science Fiction, and World Politics: Exploring Intertextual Relations", in: Jutta Weldes (ed.), *To Seek Out New Worlds: Science Fiction and World Politics*, New York: Palgrave Macmillan, p. 7; Kiersey, Nicholas J. and Neumann, Iver B. 2013: "Introduction", in: Nicholas J. Kiersey and Iver B. Neumann (eds.), *Battlestar Galactica and International Relations*, London/New York: Routledge, p. 9.
9. Weldes, J. 2003: pp. 7–8.
10. Park, Geun-hye 2016b: "Bak daetongnyeong, hanbitbudae gyeongnyeott . . . 'jinjja taeyangui huye'" [President Park Praises Hanbit Unit as 'Real Descendants of the Sun'], *Yonhap News TV*, 27 May. Available at: www.youtube.com/watch?v=_E4Fi75rOGY [Accessed 5 November 2018].
11. For a comprehensive analysis see Flamm, Patrick 2018: "No Country for Blue Helmets: South Korean National Identity on the Screen in 'Descendants of the Sun'", *The Asia-Pacific Journal: Japan Focus*, Vol. 16, Issue 11, No. 3. Available at: https://apjjf.org/2018/11/Flamm.html [Accessed 5 November 2018].
12. Lee, Myung-bak 2009: *New Year Policy Address to the Nation by President Lee Myung-bak*, 2 January.
13. Koschut, Simon 2017: "Introduction to Discourse and Emotions in International Relations", *International Studies Review*, Vol. 19, No. 3, p. 485.
14. Lee, Myung-bak 2010d: "The G20 Seoul Summit Represents the Triumph of the Korean People and the Republic", *Radio Address*, 15 November.
15. National Assembly Minutes, 2013: *Standing Committee on Foreign Affairs Session 19-321-3*, p. 7.
16. Koschut, S. 2017: p. 486.
17. Ibid.
18. Reus-Smit, Christian 2014: "Emotions and the Social", *International Theory*, Vol. 6, No. 3, p. 569.

19. Mercer, Jonathan 2014: "Feeling Like a State: Social Emotion and Identity", *International Theory*, Vol. 6, No. 3, p. 517.
20. Moon, Jae-in 2017b: *Remarks by President Moon Jae-in at the Fourth Session of the 12th G20 Summit*, 8 July.
21. Jung, Min-ho 2017: "Moon Jae-in: Son of War Refugees Rises to Power", *The Korea Times*, 9 May. Available at: www.koreatimes.co.kr/www/nation/2017/05/356_229029.html [Accessed 5 November 2018].
22. Moon, Jae-in 2017a: *Remarks by President Moon Jae-in at a Ceremony Marking the Permanent Closure of the Kori No. 1 Nuclear Reactor*, 19 June.

References

Bleiker, Roland 2001: "The Aesthetic Turn in International Political Theory", *Millenium: Journal of International Studies*, Vol. 30, No. 3, pp. 509–533.

Cooper, Andrew F. 2015: "Soft Power and the Recalibration of Middle Powers: South Korea as an East Asian Leader and Canada as the Exemplar of the Traditional Model", in: Jan Melissen and Yul Sohn (eds.), *Understanding Public Diplomacy in East Asia: Middle Powers in a Troubled Region*, New York: Palgrave Macmillan, pp. 31–50.

Denney, Steven and Friedhoff, Karl 2013: "South Korea and a New Nationalism in an Era of Strength and Prosperity", *Center for Strategic and International Studies PacNet Newsletter 75*, 7 October. Available at: https://csis-prod.s3.amazonaws.com/s3fs-public/legacy_files/files/publication/Pac1375_0.pdf [Accessed 5 November 2018].

Flamm, Patrick 2018: "No Country for Blue Helmets: South Korean National Identity on the Screen in 'Descendants of the Sun'", *The Asia-Pacific Journal: Japan Focus*, Vol. 16, Issue 11, No. 3. Available at: https://apjjf.org/2018/11/Flamm.html [Accessed 5 November 2018].

Grayson, Kyle, Davies, Matt and Philpott, Simon 2009: "Pop Goes IR? Researching the Popular Culture-World Politics Continuum", *Politics*, Vol. 29, No. 3, pp. 155–163.

Ikenberry, G. John and Mo, Jongryn 2013: *The Rise of Korean Leadership, Emerging Powers and International Order*, New York: Palgrave Macmillan.

Jung, Min-ho 2017: "Moon Jae-in: Son of War Refugees Rises to Power", *The Korea Times*, 9 May. Available at: www.koreatimes.co.kr/www/nation/2017/05/356_229029.html [Accessed 5 November 2018].

Kiersey, Nicholas J. and Neumann, Iver B. 2013: "Introduction", in: Nicholas J. Kiersey and Iver B. Neumann (eds.), *Battlestar Galactica and International Relations*, London/New York: Routledge, pp. 1–17.

Koschut, Simon 2017: "Introduction to Discourse and Emotions in International Relations", *International Studies Review*, Vol. 19, No. 3, pp. 482–487.

Lee, Myung-bak 2009: *New Year Policy Address to the Nation by President Lee Myung-bak*, 2 January.

Lee, Myung-bak 2010: "The G20 Seoul Summit Represents the Triumph of the Korean People and the Republic", *Radio Address*, 15 November.

Mercer, Jonathan 2014: "Feeling Like a State: Social Emotion and Identity", *International Theory*, Vol. 6, No. 3, pp. 515–535.

Moon, Jae-in 2017a: *Remarks by President Moon Jae-in at a Ceremony Marking the Permanent Closure of the Kori No. 1 Nuclear Reactor*, 19 June.

Moon, Jae-in 2017b: *Remarks by President Moon Jae-in at the Fourth Session of the 12th G20 Summit*, 8 July.

National Assembly Minutes 2013: *Standing Committee on Foreign Affairs Session 19-321-3*.

Park, Geun-hye 2016b: "Bak daetongnyeong, hanbitbudae gyeongnyeott . . . 'jinjja taeyangui huye'" [President Park Praises Hanbit Unit as 'Real Descendants of the Sun'], *Yonhap News TV*, 27 May. Available at: www.youtube.com/watch?v=_E4Fi75rOGY [Accessed 5 November 2018].

Park, Seo-Hyun 2013: "Changing Definitions of Sovereignty in Nineteenth-Century East Asia: Japan and Korea between China and the West", *Journal of East Asian Studies*, Vol. 13, No. 2, pp. 281–307.

Reus-Smit, Christian 2014: "Emotions and the Social", *International Theory*, Vol. 6, No. 3, pp. 568–574.

Rozman, Gilbert 2012b: "South Korean National Identity: A Six-Dimensional Analysis", in: Gilbert Rozman (ed.), *East Asian National Identities: Common Roots and Chinese Exceptionalism*, Stanford: Stanford University Press, pp. 47–72.

Snyder, Scott A. 2018: *South Korea at the Crossroads: Autonomy and Alliance in an Era of Rival Powers*, New York: Columbia University Press.

Weldes, Jutta 2003: "Popular Culture, Science Fiction, and World Politics: Exploring Intertextual Relations", in: Jutta Weldes (ed.), *To Seek Out New Worlds: Science Fiction and World Politics*, New York: Palgrave Macmillan, pp. 1–27.

Appendix

List of Inductively Retrieved Identity Elements From Both Case Studies

(Exported nodes of identity elements coded as "self-images" from 150 sources with 1,476 references in NVivo database)

Self-image	*Sources*	*References*
A country standing tall in the world.	10	12
A good neighbor to the world.	5	6
Advocate for universal values.	6	6
Being a link, node, or friend—center, not periphery.	28	34
Colonialism. Never again weak.	12	21
Colonized, because domestic elite was rotten.	1	1
Colonized, because internally weak.	1	1
Colonized, because of internal factions.	1	1
Feng Shui—flow of national spirit.	1	1
1 March 1919, as beacon of hope.	2	2
Stolen independent development.	1	1
Spirit of March 1 inherited.	1	1
Development not finished—economic, democratic, now social.	4	4
Dream of advanced nation.	32	147
Advanced country equals high gross domestic product.	5	8
Advanced means democratic.	4	7
Country in twilight zone.	2	2
Korea at the verge of becoming an advanced country.	15	32
Korea must be a place for happiness and well-being.	10	11
Rule of law and corruption.	7	8
Social safety foundation for advanced nation.	8	9
Trust and accountability.	6	8
First latecomer, now leader.	15	18
From aid recipient to donor country.	30	37
From periphery to center stage.	18	20
Global Korea.	29	70
Building a world-class nation.	6	8
Economy, trade central to Global Korea.	4	4
Korea opens for talented migrants.	4	4
Networking, international clout.	10	14
PKOs central to Global Korea.	10	10
Good global citizen.	14	19
Greater Korea.	4	4

(*Continued*)

(Continued)

Self-image	*Sources*	*References*
Military guardians of world peace.	3	3
Innovative Korea.	3	3
Korea a key player in international affairs.	10	10
Korea has always preserved dignity.	1	1
Korea as a defense industry powerhouse.	1	1
Korea as a non-Muslim country helpful.	2	2
Korea as a peaceful nation.	7	12
Korea as a (potential) green power.	39	83
Aspiring to leading as green tech powerhouse.	11	13
Aspiring to become an energy independent nation.	5	6
Exploration of polar regions.	4	11
Polar regions as treasure troves for natural resources.	1	1
Natural green leader.	1	1
Oil prices high, oil dependence.	6	7
Past carbon, future hydrogen.	3	3
ROK has no resources but ingenuity to make use of crises.	6	7
ROK as technology powerhouse.	4	4
We've caught up before, we can catch up again.	12	13
Korea at crossroads—world in crisis.	42	93
Established, advanced countries failed the world.	4	4
Korea has to take that opportunity now.	21	25
Korean adaptiveness.	1	2
Overcoming crisis through national unity.	20	26
Korea is an example of clean, good governance.	2	2
Korea not just US ally, but also mercenary.	2	3
Korean exceptionalism.	4	4
Korean rise is a surprise but a success story.	5	6
Koreans are masters of their fate/miracles.	17	26
K-pop—Korea as a cultural power.	8	14
Leadership role.	45	72
Mighty nation.	10	13
More independent foreign policy.	4	8
More proactive role in global affairs.	25	27
Next step unification.	4	4
One of select few.	23	27
Pride in miracles—museum/monument building.	2	2
Reliable ally of US.	5	9
Pro-US forces in Korea.	1	1
US alliance basis for development.	1	1
Responsible member of International community.	34	46
Rising power, new G20 power.	18	23
Korean resurgence.	2	2
ROK a liberal democracy.	7	8
ROK a major player, others want ties with us.	17	19
ROK a maritime nation.	1	1
ROK a mature country.	8	9
ROK a Newly Industrialized Country.	1	1
ROK a safe place (vis-à-vis North Korea).	3	4
ROK a sports powerhouse.	1	1
ROK an advanced economy.	1	1
ROK an advanced nation (in parts at least).	8	12

Self-image	*Sources*	*References*
ROK as a middle power.	9	10
ROK as a partner for development.	13	24
ROK has credibility (at stake).	9	11
ROK as a developed country (helping others).	37	64
Surprise story—beacon of hope for others.	11	17
ROK as a model for economy and democracy.	24	26
ROK aspiring to become model nation.	13	17
ROK a child of UN.	30	69
Paying back to UN.	18	25
UN helped ROK.	18	23
ROK country of education.	9	11
ROK country of hope for others, but not all rosy.	1	1
ROK development not only success economic but also democratic and hosting, sporting.	7	8
ROK economic, not military power.	2	2
ROK grateful, remembers past.	1	1
ROK history a miracle.	59	142
Healthy confidence—pride appropriate.	15	21
ROK *not* small and marginal anymore.	12	16
ROK was in ruins.	17	19
Small country—synonymous with war and poverty.	18	23
ROK is a small country; we should/can't do that.	1	3
ROK IT power.	3	4
ROK moral actor.	9	9
ROK not a colonizer.	1	1
ROK nuclear tech power.	7	8
ROK OECD member.	4	7
ROK power for free trade.	13	17
ROK space power.	1	1
ROK uniqueness.	3	3
Rule maker, not follower.	16	33
Exploring new paths, Korean route.	7	11
New thinking demanded.	3	4
Self-determination without too much external powers involved.	3	3
Showing pioneering spirit.	14	19
South Korean nationalism (exclusive North Korea).	2	3
Sovereign, independent nation.	3	3
Transform crisis through centrist pragmatism.	11	15
Korea and the world in harmony.	6	6
Unified Korea at center of Eurasian-Pacific Age.	9	14
Unified Korea in spirit of UN.	1	1
Widening horizon to globe.	8	9
ROK logistics and financial hub.	2	2
ROK only non-Western OECD DAC next to Japan.	5	5
Trustworthy partner.	2	3
UN Security Council seat.	1	1

Index